2009 China Interior Design Annual

2009中国室内设计年鉴（1）

《中国室内设计年鉴》编委会

辽宁科学技术出版社

目录

酒店 Hotel

售楼处 Sale Pavilion

中国式会所 Club

CONTENTS

娱乐休闲

Entertainment Leisure

老建筑改造

Renovation of old building

珞珈山国际酒店 LOGOSUN Hotel

设计单位:深圳市假日东方室内设计有限公司 / 设计:洪忠轩 / 面积:10000 m² / 主要材料:石材、LED灯、玻璃 坐落地点:武汉 / 完工时间:2008年10月 / 摄影:陈中

Design Unit:Shenzhen Holiday Oriental Interior Design Co., Ltd. / Design:Hong Zhongxuan / Area:10000 m² / Main Materials:stone、LED、glass / Location:Wuhan / Finish Time:Oct,2008 / Photography:Chen zhong

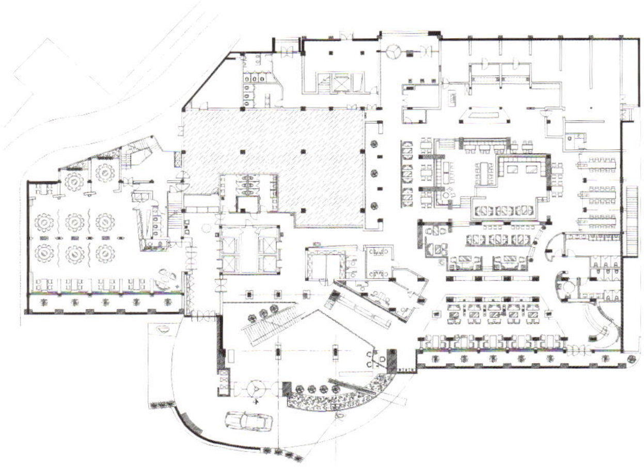

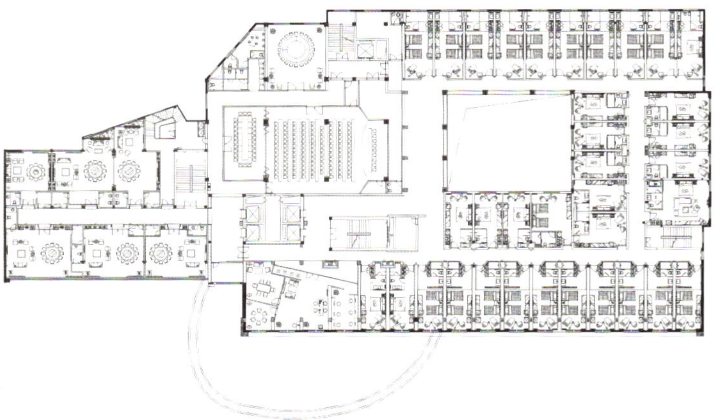

珞珈山国际酒店是纯当代概念的商务酒店，具有极简约现代特征。该酒店属于旧宾馆改造项目，位于武汉市武昌大学区。从建筑外观改造、酒店高度、通风采光等，设计师打破常规，进行全方位、多层次的构思设计，使酒店整体风格呈现出典雅融合现代时尚于一体，形成酒店独特的东方意韵和人文气质，创造五星品质的超现代人文气质的精品酒店。

珞珈山国际酒店首创全环保型酒店和生态型酒店，应用复合板材和人造木等高环保型再造材料，避免产生污染、放射和异味。大量使用高效节能灯具——LED灯，一方面有效节约了能源，推进照明技术的发展，另一方面为客户节省了不少建造成本和运营维护成本。两个具有自然采光区域的中庭，既丰富了空间的层次感，又创造了人文的自然生态环境。简约的线面结合射线型艺术灯光，不仅让有限的空间得到了视觉上的延伸，也营造了酒店不同时段的不同表情，让客人感受到一种动态的简约美。

虽然整体采用了现代简约风格，但细节上的处理绝不马虎。中庭水池的情侣雕塑、走廊的“读书”系列不锈钢艺术品、电梯厅的LED到站指示灯柱，均为洪忠轩的原创设计，充满了人性化的关怀，以及其独特的文化品位与深厚内涵。

缘于独特的酒店形象和生活方式，作为环保型酒店，该酒店成功吸引了一批新客户群，并且成功吸引海外投资，酒店的全球影响力正在与日俱增。

As a business hotel of pure modern conception, this design is characterized by the extreme simplicity and modernity. It has been designed from all aspects of its work including visual modification, high density, ventilation and day lighting etc. in order to create the first super-modern five-star exquisite hotel in China.

It is the completely environmental friendly and ecological hotel with the application of the very environmental protection type recycling materials such as composite boards and imitation woods to avoid the pollution, radiation and odors. The energy-saving LED lamps are used to save the construction, operation and maintenance costs for the customer. The atrium has not only enriched the level senses of the space, but also created a human and cultural natural ecological environment. The simple lines and surfaces in combination with the radial-style artistic light have visually extended the limited space, and have endowed the guests with a beauty of active simplicity.

简约的线面结合射线型艺术灯光，让有限的空间得到了视觉上的延伸。

The simple and concise lines and faces are combined with X-ray type artistic light so that the limited space has achieved a visual extension.

自然采光区域的中庭在LED灯的配合下时尚、独特。

The atrium in the natural day-lighting area looks fashionable and special in combination with LED lamps.

精致的细部尽显品位与内涵。
The delicate details have fully revealed the taste and meaning.
黑白纵横交织。
Black and white is crisscrossed vertically and horizontally.
艺术灯投射到楼梯上。
The artistic light is projected to the staircase.
色彩单纯、造型别致的入口阶梯。
The entrance staircase in simple color and unique model.

欧瑞锦江大酒店 Jinjiang Oriental Hotel

设计单位:YAC杨邦胜酒店设计顾问公司 / 设计:杨邦胜 / 面积:42000 m^2 / 主要材料:大理石、安利格木、桃花芯、钛咖啡板、水琉璃、艺术马赛克、艺术玻璃、墙纸、乳胶漆 / 坐落地点:重庆 / 工程造价:1.2亿元 / 完工时间:2008年12月 / 摄影:文宗博

Design Unit:YAC BANGSHENG YANG HOTEL DESIGN CONSULTING Co. / Design:Bangsheng Yang / Area:42000 m^2 / Main Materials:marble、mahogany、colored glaze、mosaic Art、art glass、wallpaper、emulsion paint / Location:Chongqin / Project Cost: RMB 120 000 000 / Finish Time:Dec,2008 / Photography:Wen Zongbo

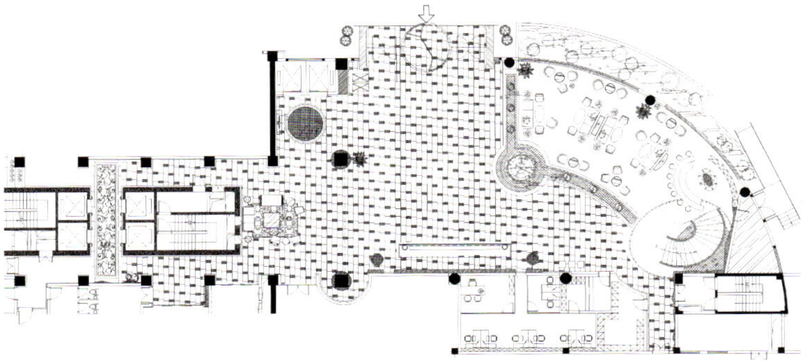

提取“花”为造型元素，抽象表达豪华富贵。
“Flower” is taken as the model element to abstractly express the luxury, wealth and rank.
珠联璧合的完美诠释。
A perfect interpretation.
光影迷离的储酒柜。
The wine cabinet in vague light and shadow.

重庆欧瑞锦江大酒店位于中国内地的保税区重庆市渝北区与江北区交汇处，地理位置得天独厚，环境优美，交通便利，由锦江国际管理公司管理的五星级城市商务酒店。设计以“花开富贵”的吉祥寓意为灵感来源，提取“花”为造型元素，贯穿整体内部空间，运用色彩的隐喻，抽象地表达出该项目豪华富贵，深沉大气的独特风格。整体空间以暖金色调和深色调相互搭配为主，金色调的雍容华贵，深色调的稳重霸气，加上花的妩媚柔美，现代质感精致的简练线条，时尚雅致的家具搭配及陈设。在华丽的光影下，缔造出一种新鲜创新的低调奢华风格。

Chongqing Ourui Jinjiang Hotel is ideally located in the only bonded area in mainland China，the intersection of Yubei Zone and Jiangbei Zone in Chongqing City. It has excellent surroundings and it is easy of access. It is a five-star urban business hotel managed by Jinjiang International Management Company.

Chongqing Ourui Jinjiang Hotel applies the auspicious symbolism of ‘Fortune Comes with Blossom’, which is the source of the design inspiration. ‘Flower’ has been picked up as the design element, and is used throughout the whole internal space. Using color as the metaphor, the project has been abstractly expressed as luxury and rich, with the characteristic of profundity and taste. The overall space is mainly coordinated by warm golden and dark colors. The prestige and nobility of golden colors, the calmness and pride of dark colors, the charm and softness of flowers, together with delicate and simplified modern lines, fashionable and elegant furniture，decorations and displays all establish a low profile, fresh yet innovative kind of luxury style under the magnificent lighting and shadow.

马赛克地面恰如繁星点缀。
The mosaic floor looks like the clusters of dotted stars.
花开富贵的寓意呈现在大理石地面上。
The meaning that fortune comes with the blossoming flowers is present on the marble floor.
极具日式餐饮风格的VIP房。
VIP room of extreme Japanese catering style.
现代质感精致的简练线条，搭配时尚雅致的家具陈设。
The delicate simple and concise lines in modern texture go with the fashionable and elegant furniture.

行政酒廊端庄优雅。

Executive lounge is decent and elegant.

宝蓝色花纹修饰的地毯流露的是一种低调的奢华。

The carpet in the sapphire blue pattern delivers a low-profiled luxury.

镜面和大理石的搭配令人视觉交错。

The arrangement of mirror with the marble generates a staggered effect in vision.

总统套房雍容华贵。

Presidential suite is natural, graceful and luxurious.

华都酒店 Waldo Hotel

设计单位:香港德坚设计公司 / 设计:陈德坚、刘永冬 / 面积:215000 m² / 主要材料:石材、玻璃 / 坐落地点:澳门 / 完工时间:2008年10月 / 摄影:陈德坚

Design Unit:Kinney chan and associates / Design:Kinney chan、Liu Yongdong / Area:215000 m² / Main Materials:stone、glass / Location:Macao / Finish Time:Oct,2008 / Photography:Kinney chan

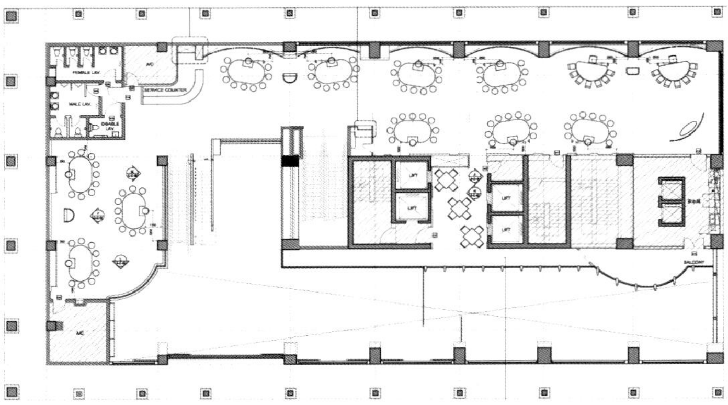

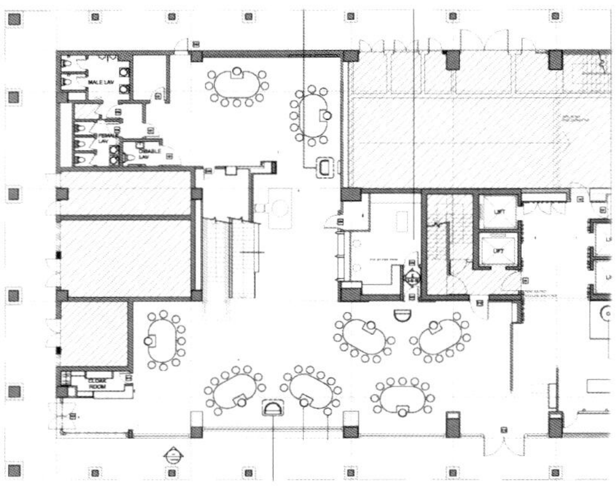

新口岸的华都酒店本属传统类型，这次设计师将华都酒店重新包装，以拉斯维加斯Boutique Hotel概念设计，变身成澳门第一间设计新颖、富有时代感的精品酒店。Boutique Hotel希望给住客营造一份家的感觉，因此在装潢上会较温暖和谐，亦会别致地在房内分隔出客厅等各部分，有些甚至设有厨房。华都的百多间客房中，有一半是设计很有性格的主题套房。

重新设计的华都酒店创意十足，突显一个“潮”字。外观，采用LED显示标志，两颗“红、绿”的圆点标记设于外墙，其高亮度、对比度、图像质量简约而醒目。大堂，椭圆金属天花悬挂在电梯大堂的中间，接待柜台后面，特别设计的紫色洗墙灯，神秘而暗哑。酒店入口处一层，设有雪茄吧和红葡萄酒吧，周围铺着冷峻的大理石。空间流线明晰，淡黄、玫瑰红牵引着视觉。流线型天花装饰及地毯图案，既美观又有引路作用。内部黝黑皮革家具和低调的照明，现代而优雅，木制的桌面焕发出热情和黄金的氛围。玻璃墙板空间中心安装电线，射灯照在玻璃墙板上，营造出墙壁戏剧性的效果，同时增添了现场钢琴表演的气氛。赌场空间，顶部圆形天花板模式和地毯形状相互呼应，每张桌旁玻璃吸顶灯，与魅力时尚的家具组合，无不彰显富贵与奢华。

而餐厅分为西餐和中餐，设计风格平静和简单，走廊在昏暗灯光中品味舒适和轻松的气氛。黑色的大理石地面反射在镜面，勾勒出形态优美的曲线。顶部利用大面积的镜子，扩展了空间的广阔。硕大的玻璃屏风，营造虚实气氛。黑白的花纹石柱，体现餐厅的异样艺术与文化。靠近酒店入口的“Perle”Bar，在设计中强调了酒吧周围柜台顶部和墙壁，外观增加了抛光度，入口处装饰闪烁和运动银色珠子，与暗淡的灯光相结合，伴着音乐，让疲惫的身心得到放松。

客房、酒店共有162间，其中两间是顶楼总统套房，四种主题房间：Glamour、Pure、Comfort及Hip，分布于16~18楼，其中最惊喜是银黑幽暗酒红地毡的Glamour，面积约500m²，名副其实的“格”型设计，在顶部上方“套”入一个相架，大有画龙点睛效果。至于白色为主调的Pure、大自然色彩的Comfort及波浪形衣柜的Hip，都让客人耳目一新，印象难忘。

As Macau has fast become the largest and the most popular destination leisure, entertainment and gaming marketing worldwide, Waldo Hotel has decided to reposition the brand and re-renovate itself with a more attractive image to ultimately be competitive in the marketplace.

Located in the heart of Macau, Waldo Hotel is a provocative hotel offers the perfect retreat while also providing gambling services. Waldo undertakes the upgrade and re-bands the property to a Hip and trendy hotel. The overall goal was to brighten up both the guestrooms and the public spaces, including the lobby and restaurant. The interior design theme of Waldo Hotel is to create a comfortable, exciting and unique atmosphere.

The hotel facade features an impressive LED logo display, its high brightness and contrast ratio, and the stunning image quality, clearing presenting the trendy image of Waldo.

The presentation of main lobby is in keeping with the hip and chic image. A very nice modern and trendy color schemes such as noble and burgundy have been applied into the area. To bring out the hip ambiance, an oval metal ceiling feature has been hung in the middle of the lift lobby. Behind the reception counter, there is a silver wire feature wall with purple wash light. All furniture is in simple form yet natural and cozy.

The exciting and glamorous theme is carried throughout the interior design of the hotel casino. The interior design was similarly fashionable, with hip circular patterns on the ceiling and patterns on the carpet. The column glassy ceiling lamps, as well as the suede upholstery furniture successfully warm up the atmosphere and create a customer friendly environment.

There are four type of themed guest room inside the hotel: Pure, Comfort, Hip and Glamorous. The decoration of “Pure” is simple, white and minimal but still inviting and comfortable. The room “Comfort” however is decorated in natural warm tone for a cozy and elegant look. Soft touch such as the carpet imparts a vivid modern flavor. The Hip room is completely chic and hip with unique pieces of modern furniture. Vivacious use of material and fabric in the interiors make it a sigh to behold. In "Glamorous", the design is fascinating and sumptuously appointed. Tasteful furnishings and color tones make it even more luxurious.

The western style restaurant Verde is another trendy dining area of the hotel. The interior main setting is calm and unsophisticated, as well as dim lightings in the corridor generates a cozy and relaxing atmosphere. The floor and counter bar are made of black marble. Several illuminated tinted mirror columns with patterns of elegant curves carrying out a vivid effect. With the use of mirror on the ceiling, height of the space has been wisely widened. Dining in this discreet elegance restaurant is an experience to be cherished.

The Perle is an exclusive bar located just off the hotel entrance. The bar offers a classy and charming atmosphere, with fashionable d é cors, opulent surrounding, together with its great selection of wine and drinks make this upscale venue a perfect location for a relaxing night out.

To create a glamorous effect, designer adds a polished and shiny appearance into the bar by surrounding the counter top and the wall with white shiny mother of pearls. The entrance is decorated by shimmering and motional silver beads. Combining subdued lighting, as well as comfortable silver leafed arm chair and others impeccably stylish furniture, the design successfully creates a charming and glamorous visual effects.

Located on the 1st floor of the hotel, Kuro is an upscale cigar and red wine lounge aims to provide custom-tailored service to distinguished guests. The interior is modern and classy, with dark leather furniture and low key lightings, designer attempts to keep in with the cigar bar theme. The wooden tabletops radiate warmth and golden ambience, while heavy dark curtains keep out glaring neon lights from outside. A feature wall panel standing in the heart of the space is created with wires, with spot lights shooting from the floor, the wall carries out a dramatic effect in this calm environment. With live piano performances add to the ambience, Kuro will be an ideal location to enjoy a remarkably high standard and stylish drinking and smoking experience.

Jaune is an innovative Chinese cuisines restaurant served in an impressively chic setting. It offers Chinese fusion cuisine, and it has been carefully created to accommodate contemporary tastes in the design. Several accessories and decors give a modern touch. This stylish yet harmonious design of the interior made it an exciting place.

设计中强调了酒吧周围柜台顶部和墙壁，外观增加了抛光度。

The design has emphasized the top of the counter and wall around the bar and increase of visual polishing.

空间流线明晰，淡黄、玫瑰红牵引着视觉。
The space has clear streamlines. The canary and rosy red attract your eyes.
接待柜台后面，特别设计的紫色洗墙灯，神秘而暗哑。
The specially design purple wall washing lamp behind the receptionist desk looks mysterious and shadowy.
酒店入口处一层，周围铺着冷峻的大理石。
The floor at the entrance of the hotel is covered by the cool marble.

黑白的花纹石柱，体现餐厅的异样艺术与文化。

The black and white patterned stone columns have shown the different art and culture of the restaurant.

卫生间设计风格平静、简单，在浪漫的灯光中品味舒适和轻松的气氛。

The washroom is designed calm and simple. You can enjoy the comfortable and relaxing atmosphere in the romantic light.

银黑幽暗酒红地毯的Glamour，是名副其实的“格”型设计，顶部上方“套”入一个相架，大有画龙点睛效果。

The glamour of the silver black dim claret carpet is the real design of “grid” model. The top is “encased” with a photo frame to make the finishing point.

瑞亚·臻品酒店（梅园路店） RIA-Perfection Hotel(Mei-yuan road's)

设计单位:美国PDA室内设计事务所 / 设计:裴晓军 / 面积:6000 m² / 主要材料:诺娃米黄石、玫瑰木、壁纸、布艺、钢材 / 坐落地点:上海 / 工程造价:700万元 / 完工时间:2008年10月 / 摄影:贾方

Design Unit:USA PDA DESIGN / Design:Pei Xiaojun / Area:6000 m² / Main Materials:rosewood、wallpaper、cloth、steel / Location: Shanghai / Project Cost:RMB 7 000 000 / Finish Time:Oct,2008 / Photography:Jia fang

自从有了精品酒店这样的概念后，全球的旅馆开始仿效，一些作品视为当今的经典。大型酒店、小型酒店、酒店的品种随客户的需求变化而变化，特别在今天中小型酒店在市场冲击和打压下开始突围，寻求生路。于是林林总总的酒店应运而生，琳琅满目，令人不可想象的蜕变……

瑞亚·臻品酒店前身是一个经济型旅馆，再前身是工厂的加工车间。由于建筑的多次改造增扩，构造柱没有规律，也造成空间无序，设计师团队如何重建基础设施和酒店系统，达到业主的要求。于是才有了本案。

Since these years have top quality hotel idea, so all the hotels are imitative of this in the world, and many productions were classical. All kinds of them have change for the client, especial to-day in the impact of market pressure, like medium or small hotel look after new road, so all different of them have been live and bring a big transformation.

RIA-Perfection Hotel was an economic hotel before, and a long time ago it was a factory. Because this building had been many reconstruct and worked the structure and space no rule no order. So designer team began to rebuild this hotel base and system for the client, and the case was born.

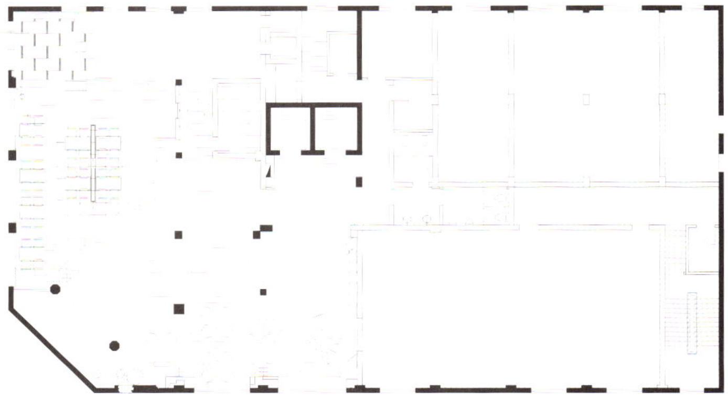

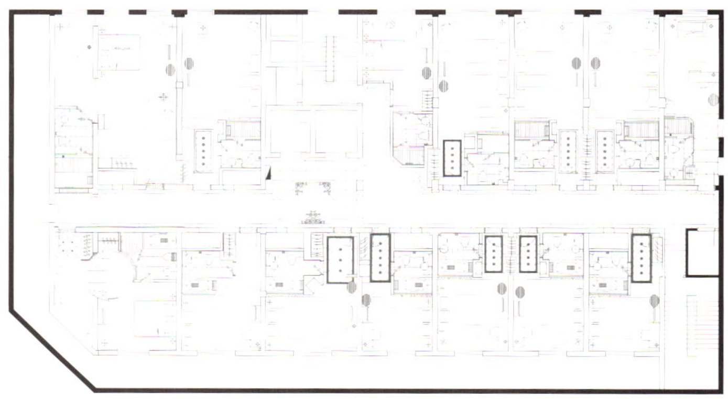

地毯、软包、家具，力求色彩上极致统一。

The carpet, soft packs and furniture try to achieve a perfect unification on the colors.

玫瑰木与米黄石双剑合璧未尝不是品质的升华。

The perfect combination of rosy wood and off-white stone is certainly the enhancement of the quality.

饰有花纹的玫瑰红灯饰柔化了冷峻的大理石。

The patterned rosy light decoration has softened the cool marble.

秩序井然的客房过道。

The orderly guest room aisle.

叶瓣镂空屏风平添了大堂些许复古中式味道。

The leaf-carved screen adds certain antique Chinese charm to the lobby.

黑色烤漆玻璃和镜子的搭配助长了自助西餐的冷郁低调。

The black lacquered glass goes with the mirror to strengthen the cool fragrance and low profile of the buffets.

欲闭还合的客房功能齐全。

The guest room that is close in order to close has complete functions.

点到即止的照明交由客人更多的适意。
The lighting to the proper extent provides more comforts to the guests.
曲线流动的墙、顶面赋予空间灵动的气息。
The curve-flowing wall and ceiling have endowed the space with bright atmosphere.

太子城假日酒店 Prince City Holiday Inn

设计单位:YAC杨邦胜酒店设计顾问公司 / 设计:杨邦胜 / 面积:40000 m² / 主要材料:大理石、白影木、安利格木、泰柚、灰镜 / 坐落地点:本溪 / 工程造价:2亿元 / 完工时间:2008年11月 / 摄影:文宗博

Design Unit:YAC BANGSHENG YANG HOTEL DESIGN CONSULTING Co. / Design:Bangsheng Yang / Area:40000 m² / Main Materials:marble、wood、teak、neutral、density filter / Location:Benxi / Project Cost: RMB 200 000 000 / Finish Time:Nov,2008 / Photography:Wen Zongbo

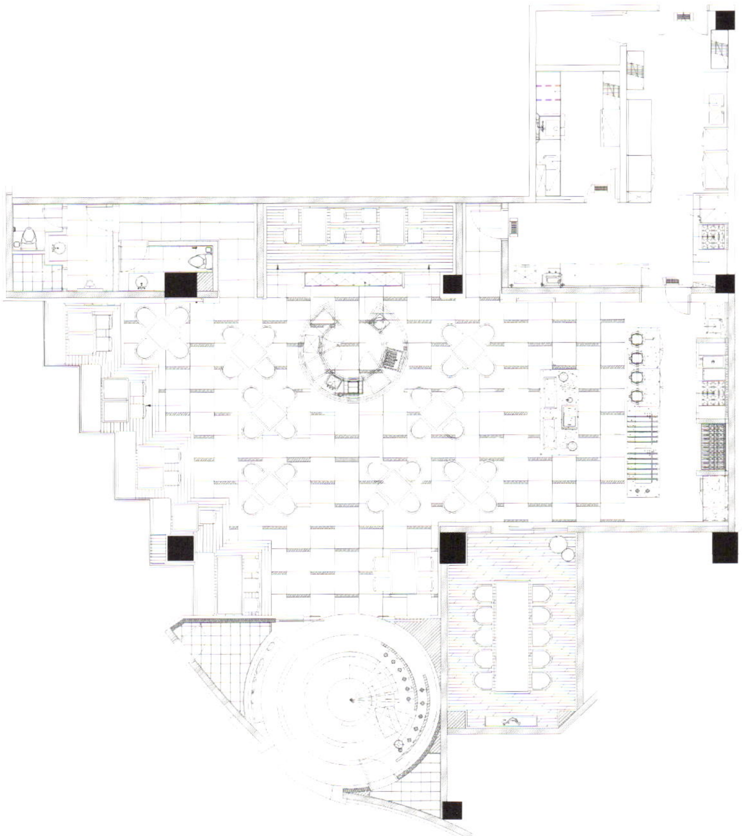

辽宁本溪是一个以枫叶而闻名的美丽城市，本溪太子城假日酒店坐落在太子河畔。整个酒店的设计简洁而时尚，以枫叶为主题，将枫叶的元素提炼、抽象化，从公共区域到客房始终将之贯穿其中，并结合了中国风水“金、木、水、火、土”的理念。酒店空间分为五种色彩：中餐以“金”为贵；全日餐厅体现出了“木”的绿色生态气息；康乐区以“水”为主；巴西烧烤餐厅将红色“火”的理念发挥到极致；浅色的会议空间则体现出了“土”的庄重；五色俱用，体现出酒店的质朴大方，而又不失现代时尚。空间和谐而舒适。

Liaoning Benxi is a beautiful city known for its maple leaf. Benxi Prince City Holiday Inn is located at the Prince riverside. The design of the hotel is simple and modern with maple leaf as its theme. The abstracted elements of the maple leaf are applied from public area to the guest rooms. Combining the concept of five Chinese Fengshui elements, namely gold, wood, water, fire and earth, the hotel space is mainly of five colors: Chinese restaurant uses noble gold; all-day ecological Japanese restaurant reflects the green ambience of ‘wood’; recreation area uses ‘water’ mainly; Brazilian barbecue restaurant brings into play the red ‘fire’ to its perfection; the light-colored meeting space reflects the dignity of ‘earth’. The proper combination of the five colors reflects the rusticity and decency of the hotel without losing modernity and vogue, which is also harmonious and comfortable.

公共休息区犹如小型居家客厅，温馨甜蜜。
The public resting area is just like a small living room, warm and happy.
线帘围合出的小小世界。
A small world enclosed by the string curtain.

各式工艺品的陈列让人误以为走进某家私人博物馆。
The display of various artworks has mistaken the people of entering a private museum.
全日餐厅体现出了“木”的绿色生态气息。
The all-day restaurant has shown the green ecologic atmosphere of “wood”.
红与黑的经典永远不过时。
The classic red and black is never out of date.
巴西烧烤餐厅将红色“火”的理念发挥到极致。
Brazilian BBQ restaurant has developed the philosophy of red “fire” to the greatest extent.

即使过道一隅，亦能窥视出整个酒店的品位。
The taste of the whole hotel can been found even at one corner of the corridor.
木色的运用体现酒店的质朴大方，而又不失现代时尚。
The application of the wood color has reflected the simple, grand and modern fashion of the hotel.
除了酒店主题元素枫叶的运用，客房几乎未加任何装饰。
The guest rooms have hardly any decoration expect the application of the thematic element of the hotel – maple leaf.
单人间套房和谐而舒适。
The single suite is harmonious and comfortable.

太和紫金大酒店 Taihe Zijin Grand Hotel

设计单位:江苏省海岳酒店设计顾问有限公司 / 设计:姜湘岳、吴海燕 / 参与设计:卞玲 / 面积:11000 m^2 / 主要材料:布纹格地砖、布纹格大理石、液晶面板 / 坐落地点:南京 / 工程造价:6700万元 / 完工时间:2008年12月 / 摄影:莫尚勤

Design Unit:Jiangsu Haiyue Hotel Design Consulting Co., Ltd. / Design:Jiang Xiangyue、Wu Haiyan / Participatory Design:Bian ling / Area:11000 m^2 / Main Materials:floor tile、marble、TFT-LCD / Location: Nanjing / Project Cost: RMB 67 000 000 / Finish Time:Dec,2008 / Photography:Steve Mok

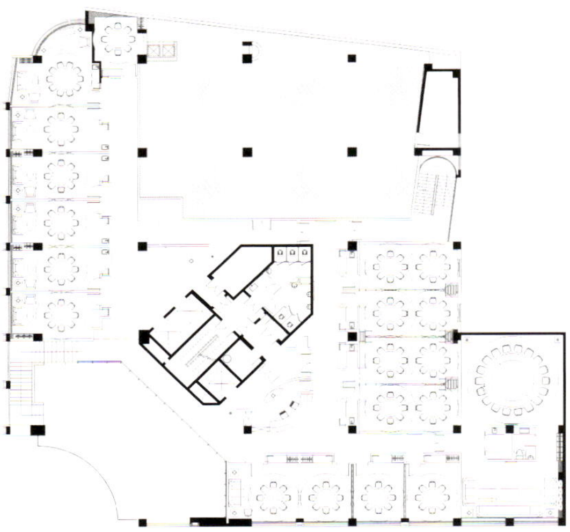

二楼包厢入口处。

The entrance to VIP rooms.

拾级而上，黑色大理石在灯光的辉映下极具品质。

When walking up the steps the black marble looks extremely superior under the reflection of the light.

本案位于一个非常时尚的街区对面，因此设计理念是要做一个融合很多时尚元素的精品型酒店。

大堂——设计了一个横跨的封闭式廊桥，让往来和进入的客人没有目光上的碰撞。入口正对面放置了两匹黑色的马，让人有一种田园牧场的感觉。

西餐厅——色调做得很灰，用白梨木雕刻出以羊为主题的艺术品有序列地放置在就餐位置的旁边，一来作了遮挡，二来表达乡村气息。地面采用黑色有条纹格的石材，这一切让人感觉很怀旧也很时尚。

由于建筑标准层的层高只有3m，对设计的制约非常大，很多的功能都不能满足。主要体现在两个区域，一个是客房，一个是宴会厅。所以在设计时我们做了大量的工作，希望能通过我们的设计来改变这种状况。

在客房的设计上，我们采用了非常单纯清新的浅色调，为了增加高度，我们把设备风管进行了重新的改造，减少了新风管对走道的压抑，使走道的高度能尽可能提升。另外将家具和空间的颜色拉开对比，这样可使空间更为明朗。

考虑到酒店的功能配套，必须要做一个宴会厅，但是宴会厅的梁底完成面只能做到2.2m，这给我们的设计带来了很大的难度。为了实现这个功能，我们在设计时将其顶部全部采用镜面，墙面强调竖线条，这样就让空间瘦起来、拔高起来、反射起来，从而视觉上把整个高度复制了一倍，等于让客人看到的层高是两层高，把压抑感和闷的感觉完全消除掉了。虽然最低点只有2.2m，但是给客人的感觉一点都不压抑，非常的透气，来参加宴会的客人也都很喜欢这样的空间。

酒店大堂前台的背景专门请日本的墙纸艺术家做了莲花的图案，象征和谐，美好。“太和”——太平、和谐，又象征着当今的和谐社会，是一种对美好愿望的表达。

This project is located opposite to a very fashion block. Thus, our design philosophy is tended to build a fashionable and exquisite hotel integrated with many fashion elements.

Lobby – a cross enclosed corridor bridge is designed to prevent looking at each other of the coming and going guests. The entrance of the lobby is right opposite to the two black horses, which has demonstrated a very fresh and natural feeling of a garden city and pasture.

Western Restaurant – the colors are made quite grayish. The arts carved by white pearl wood with sheep as the subject are placed beside the dining tables in order, which on the one hand, can be used as a block, on the other hand, can be used to deliver a countryside atmosphere. . The ground is covered with the black strip-and-grid stone materials, which looks reminiscent and fashionable.

As standard storey of the building is only 3m and many functions can not be met, mainly reflected in two areas: on is the guest room and the other is the banquet hall. Therefore, we have made a great quantity of work and hope to change this situation through our design.

In the design of the guest rooms, we have applied very simple and fresh light colors. To increase the height, we have re-modified equipment ducts in order to reduce the constraints of new ducts on the corridor and uplift the height of the corridor as high as possible. Furthermore, the colors of the furniture and space are contrasted greatly so that the space looks brighter.

In consideration of the functional mach for the hotel, a banquet hall shall be made. But the finish of the beam bottom in the banquet hall can only be made up to 2.2m, which has brought great difficulties on the design. To realize this function, we have applied the mirror to the whole ceiling while the wall emphasizes on the vertical lines so that the space gets thinner and reflected so that the whole height is copied visually, which means that the floor height the guests can see is two stories and completely eliminate the constraints. Although the lowest point is only 2.2m, the feeling to the guests is not oppressed at all and very ventilate. The guests to attend the banquets also like such space.

The background of the reception desk in the lobby is made into the lotus pattern by a Japanese wall paper artist, which symbolizes the harmony and goodliness. “Taihe” – peaceful and harmonious, also symbolizes the current harmonious society and is an expression to the good wills.

和軒
CHINESE CUISINE

和轩中国料理包间入口处。

The entrance to VIP rooms of Hexuan Chinese Food VIP rooms.

自助料理区。

Self-served Food Area.

线条明快，动线流畅的公共区域。

Public Area with bright lines and smooth active lines.

入口正对面放置了两匹黑色的马，人有一种田园牧场的感觉。

The entrance of the lobby is right opposite to the two black horses, which has demonstrated a very fresh and natural feeling of a garden city and pasture.

用红色的小果子灌满的罐子组合成的墙壁别有一番韵味。
The wall combined by the pots full of red fruits looks extremely charming.
白梨木雕刻出以羊为主题的艺术品有序列的放置在就餐位置的旁边。
The arts carved out of white pearl wood with sheep as the subject are placed beside the dining tables in order.
点点灯光排列于餐桌上方，精致、温馨。
The dotted row upon row of lights is lined up above the dining tables, delicate and warm.
客房设计上采用非常单纯清新的浅色调。
The design of guest rooms has applied the quite simple and fresh light colors.
家具和空间的颜色拉开对比，使空间更为明朗。
The distinction and contrast between the colors of the furniture and space has made the space brighter.

丽思·卡尔顿酒店 Ritz-Carlton Hotel

设计单位:深圳市姜峰室内设计有限公司 / 设计:姜峰、陈礼庆、袁晓云 / 面积:30000 m^2 / 主要材料:非洲米黄石、七彩玉石、金黄洞石、金龙米黄石、紫点金麻石、白玉石、酸枝木、古铜色拉丝不锈钢 / 坐落地点:深圳 / 工程造价:10亿元 / 完工时间:2008年3月

Design Unit:J&A Interior Design CO.,LTD / Design:Jiang feng、Chen Liqing、Yuan Xiaoyun / Area:30000 m^2 / Main Materials:stone、travertine、dalbergia cochinchinensis、brushed stainless steel / Project Cost: RMB 1 000 000 000 / Finish Time:Mar,2008

丽思·卡尔顿酒店位于深圳市福田中心区，邻近深圳会展中心，共24层，拥有273间豪华客房，设有中餐厅、意大利餐厅、咖啡厅、500人宴会厅、酒吧等，是深圳目前最奢华的白金五星级酒店。

酒店设计风格摒弃丽思·卡尔顿惯用的西方古典豪华风格，而从中国传统文化中提炼中式元素，融入现代设计手法，形成独具中国文化特征的现代酒店风格。在空间营造方面，三层高的大堂，气派非凡，一层设有大堂吧、丽思吧，通过家具的摆设，通透的屏风隔断，空间错落有致，互相渗透，内容丰富而没有拥挤感，空间宽敞高大而没有空洞感，营造了豪华舒适的酒店大堂氛围。在装饰题材的运用上，提炼出独具中国人文特色的“梅”、“兰”、“菊”、“竹”主题，运用现代工艺，赋予其新的形式。

Shenzhen Ritz-Carlton Hotel is located in the center of Futian District, Shenzhen, the neighbor of Shenzhen Convention and Exhibition Centre. The hotel is a total of 24 stories. It has 273 luxury guest rooms, Chinese restaurants, Italian restaurants, a coffee house, a banquet room for 500 people and bars. It is the most luxurious five-star platinum hotel in Shenzhen.

The design abandons western luxury classical styles consistently used by Ritz-Carlton; instead, it extracts Chinese elements from traditional Chinese culture, blended with modern design arts to form a modern hotel with unique Chinese cultural characteristics. In terms of space design, the three-story lobby looks extremely splendid. The ground floor is equipped with lobby lounges and Ritz bars. The space seems orderly and transparent, rich but not crowded, specious but not empty, creating a luxurious and comfortable atmosphere in the hotel lobby. In using of decoration themes, designers extract the Chinese cultural elements "Plum Blossom", "Orchid", "Chrysanthemum" and "Bamboo" as the theme, echoed with the modern technology to endow them a new form.

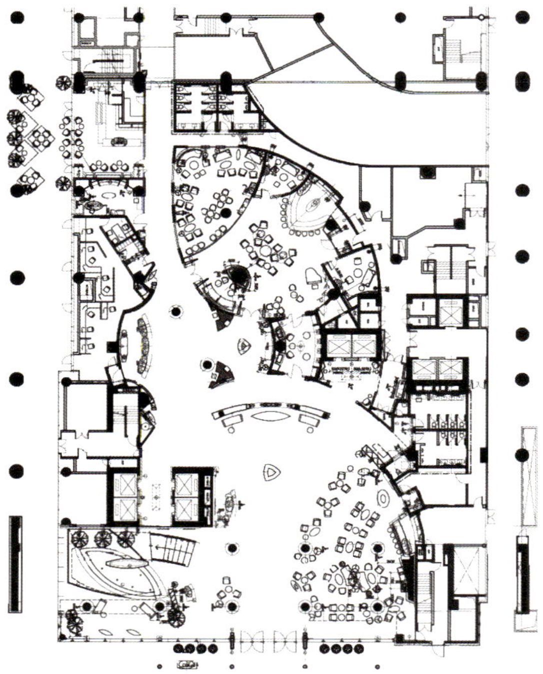

富丽堂皇的吧台。
Magnificent Bar Counter.
三层高的大堂，气派非凡。
The three-level lobby looks grand and outstanding.

通过家具的摆设，通透的屏风隔断，形成错落有致的空间。
A proper and orderly space is arranged through the placement of the furniture and the transparent screen partitions.
豪华包间大气磅礴。
The luxurious VIP room looks grand and majestic.
空间内容丰富而没有拥挤感。
The space is vivid without any sense of crowdedness.

高低层的形态令空间灵动。

The forms of high and low levels make the space bright and charming.

内涵丰富的细部。

The meaningful details.

硬装的暖黄与软饰的冷峻相得益彰。

The warm yellow of the fixed decoration and the coolness of the soft decorations complement each other brilliantly.

"三"的韵律 Rhythm of "Three"

锦江之星–宁波柳汀街店

设计单位:上海泓叶室内设计咨询有限公司 / 面积:1000 m² / 坐落地点:宁波 / 完工时间:2009年3月 / 摄影:叶铮

Design Unit: HYID / Area:1000 m² / Location:Ningbo / Finish Time:Mar,2009 / Photography:Ye zheng

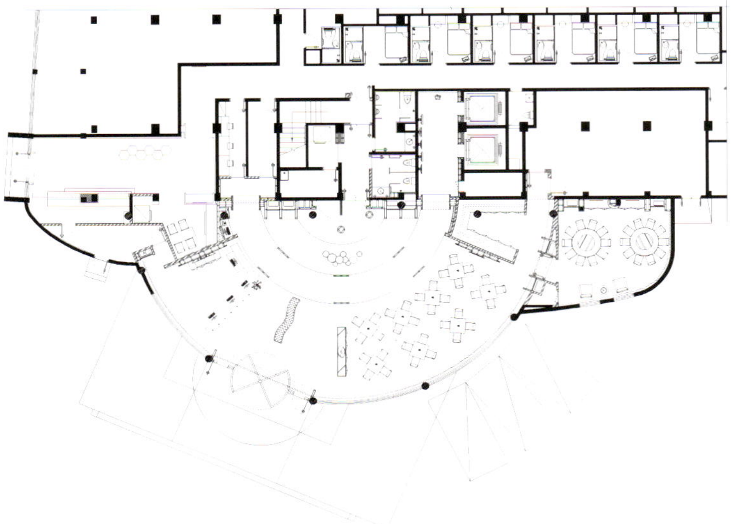

位于宁波市中心繁华地带的柳汀街上，有一座在丁字形路口被改建的酒店，其建筑面积约为9000m²，共计客房为250间。由于地形关系，建筑平面呈扇形展开，这首先成为室内设计的切入点。以富有严谨秩序感的弧型布置，构成该酒店内部空间的第一印象。其次，以"三"的倍数构成1：3竖向界面的矩形语言，形成富有节奏韵律的抽象关系与比例，使该项目含有极强的数字逻辑美感。而陈设设计又是本案的一大特色，无论是圆几群，还是漂浮的灯具设计，抑或是那些飞跃在高耸框架内的亮银色蝴蝶群，无疑都为室内空间平添了动人的印象。

Located at Liuting Street in the busy zone of Ningbo downtown area, there is a renovated hotel at the T-crossing with a floor area of about 9,000m2 and a total number of 250 various guest rooms. The building layout outspreads in a sector due to the landform, which first becomes the approach to this interior design. The first impression on the interior space of this space is the arc arrangement in a precise and systematic way. Secondly, a rectangular language in a vertical interface with a ratio of 1:3 is constituted by the multiple of "three" to form a rhythmic abstract relationship and ratio so that this project has an extremely strong digital and logic beauty. Furthermore, the design of the displays is another big feature of this case. No matter the round shapes, the design of floating lighting fixtures, or those bright silver butterfly throngs flying in the towering frame, doubtlessly, they naturally bring the attractive impression to this interior space.

以富有严谨秩序感的弧形布置，构成该酒店内部空间的第一印象。
The first impression on the interior space of this hotel is the arc arrangement in a precise and systematic way.
纱帘恰似一泻千里的瀑布述说着诗意的生活。
The gauze curtain is just the bold flowing fall the depicting the poetic life.

不同用光手法的纯熟运用释放空间婉约的魅力。

The skilled application of the different ways of light management delivers the graceful and restrained charm of the space.

高低错落的圆几群如音符般奏响柔美的乐章。

The round shapes in different levels look like the music tones playing the gentle music.

白色桌椅搭配浅蓝桌布，再配上一朵鲜花，精致浪漫的生活理念悄然流淌。

The white table and chairs in arrangement with the Cambridge blue table cloth together with a fresh flower is quietly delivering a delicate and romantic living philosophy.

飞跃在高耸框架内的亮银色蝴蝶群，为空间平添了动人的印象。

The brilliant silver butterfly groups flying in the towering frame have added the affecting impression to the space.

风雅姑苏 Elegant Soochow

锦江之星–苏州人民南路店

设计单位:上海泓叶室内设计咨询有限公司 / 面积:600 m² / 坐落地点:苏州 / 完工时间:2008年12月 / 摄影:叶铮

Design Unit:HYID / Area:600 m² / Location: Suzhou / Finish Time:Dec,2008 / Photography:Ye zheng

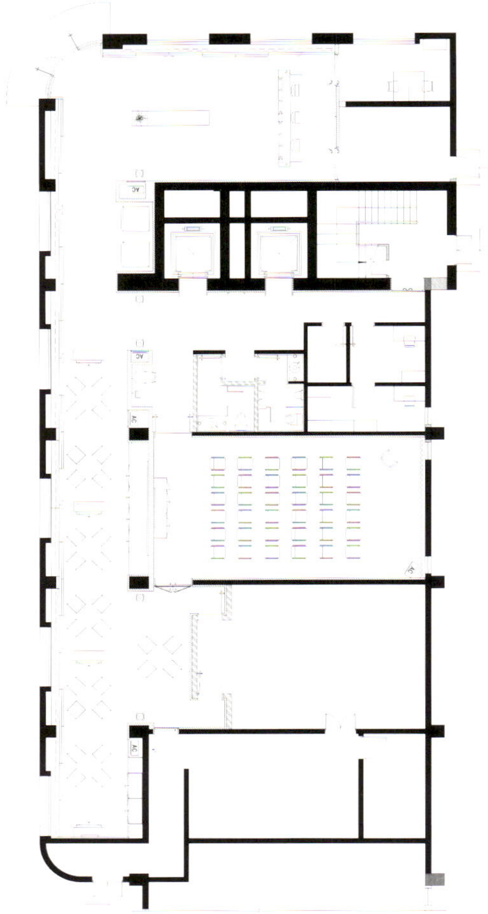

酒店位于苏州人民南路，历来的素雅与人文气息成为设计中无法回避的传统。为营造安宁的气氛，设计在邻街的落地玻璃内侧安置了可移动的落地花格窗，既隔绝来自外界的干扰，又不失室内采光，花格背后的白冰绸，透光而不透明，使室内弥漫着柔和宁静的感觉。晚上，灯光同样透射过朦胧的白冰绸，衬托着精致的传统花格。为使设计更具现代感，选择了具有金属光泽的喷漆。同样的金属材质还出现在家具陈设中，以获得时尚的苏州印象。室内空间在色调上加重了顶与墙界面。如此，既可强调透光的花格陈设，又可产生室内的诗意。苏州店设计，旨在向传统文人的儒雅内敛作出当代的诠释。

The hotel is located in South Renmin Road in the busy downtown area of Suzhou. The elegant and human cultural atmosphere has always been the unavoidable tradition in the design. To create a peaceful environment, the movable sliding floor pane windows are arranged inside the floor glass beside the street, which can not only prevent the external interferences, but also can be used for interior day lighting. The white icy silk behind the pane, photic but not transparent, has filled the interior with a soft and peaceful feeling. At night, the light also transmits through the dim white icy silk to set off the delicate traditional pane. The spray paint of metallic gloss is selected to make the design look more modern. The same

Metal materials also appear on the furniture to achieve a fashionable Suzhou impression. For its interior space, the colors for the ceiling and wall interfaces are stressed so that it can not only emphasize the photic pane display, but also can generate the interior poetry. The design of Suzhou hotel tends to make a contemporary interpretation on the genteelness and restraint of the traditional scholars.

花格背后的白冰绸，使室内弥漫着柔和宁静的感觉。
The white icy silk behind the panes has permeated the interior with a gentle and peaceful feeling.
室内空间在色调上加重了顶与墙界面。
The interior space has stressed the ceiling and wall interfaces in color.

灯光透射，衬托着精致的传统花格。
The light transmission sets off the delicate traditional panes.
可移动的落地花格窗，既隔绝来自外界的干扰，又不失室内采光。
The movable paned floor window not only isolates the external interferences, but also provides the interior day lighting.
素雅与人文气息成为设计中无法回避的传统。
The simple, elegant and human cultural atmosphere has become the unavoidable tradition in the design.

"圆"的叠加
锦江之星–潍坊东风西路店

Overlapping of "Rounds"

设计单位:上海泓叶室内设计咨询有限公司 / 面积:700 m^2 / 坐落地点:潍坊 / 完工时间:2009年1月 / 摄影:叶铮

Design Unit:HYID / Area:700 m^2 / Location:Weifang / Finish Time:Jan,2009 / Photography:Ye zheng

山东潍坊，一个以风筝闻名的地方。印象中，风筝那艳丽的色彩浮现于眼前。其中，以红色尤为突出。设计恰是以这样超现实的跨度，完成了空间界面的概念定位。各式大小不同的圆形相互叠加，在黑色钢琴漆的衬托下，构成深浅不等的浮雕感。圆成为设计的主题元素。同时，在黑、银与冷艳的红色之间，形成色彩的基调。而在材质组织上，则以水性与油性的对比建立起亚光与反光的材料关系。整个设计，以装饰图案为表现语言，使空间氛围异常时尚、冷艳，并求妩媚有力、柔中济刚的两极统一。

Weifang, Shangdong is famous for the kites. In our impression, the gorgeous colors of the kites are emerging in front of us, among which red is extremely prominent. This design has completed a conceptual orientation of the space interface by such super–realistic span. The various types of rounds in different sizes are overlapping with each other and have formed a sense of reliefo in irregular shades under the setoff of the black piano lacquer. The round has become the thematic element of the design and further formed the basic color of this case among black, silver and quiet elegant and magnificent red. In regard to its material arrangement, the contrast between water and oil natures is applied to set up a relationship between the matte and luminescent materials. The whole design has taken the decorative patterns as the expressive language to make the space atmosphere extremely fashionable, cool and magnificent and reach a unification of two extremities – the charm and power as well as softness and firmness.

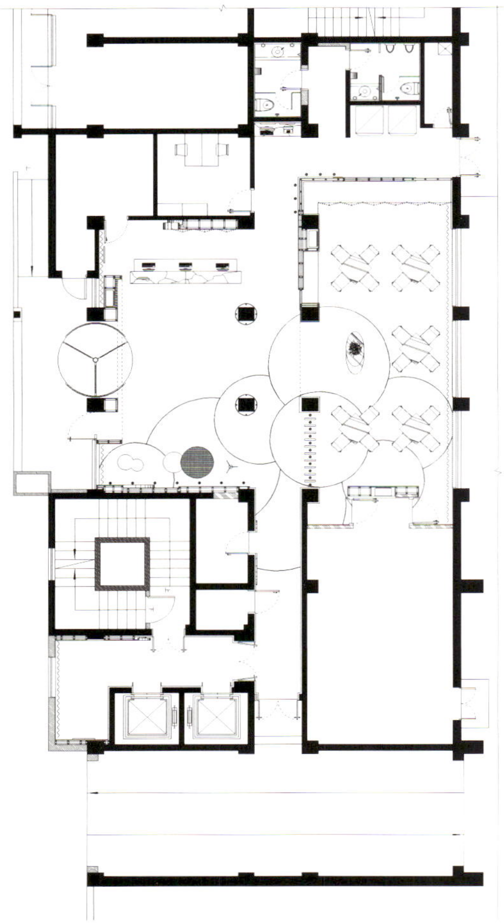

各式大小不同的圆形相互叠加。
The rounds in different sizes are overlapping with each other.
突出的红色令空间妩媚有力。
The prominent red has made the space charming and magnificent.

在黑色钢琴漆的衬托下，圆形构成深浅不等的浮雕感。

The rounds have formed a sense of relievo in irregular shades under the setoff of the black piano lacquer.

水性与油性的对比建立起亚光与反光的材料关系。

The contrast between water and oil natures is applied to set up a relationship between the matte and luminescent materials.

在黑、银与冷艳的红色之间，形成了本案色彩的基调。

It has formed the basic color of this design among black, silver as well as the quiet elegant and magnificent red.

视幻空间 Visional Space

锦江之星-武汉紫阳路店

设计单位:上海泓叶室内设计咨询有限公司 / 面积:500 m² / 坐落地点:武汉 / 完工时间:2009年4月 / 摄影:叶铮

Design Unit:HYID / Area:500 m² / Location:Wuhan / Finish Time:Apr,2009 / Photography:Ye zheng

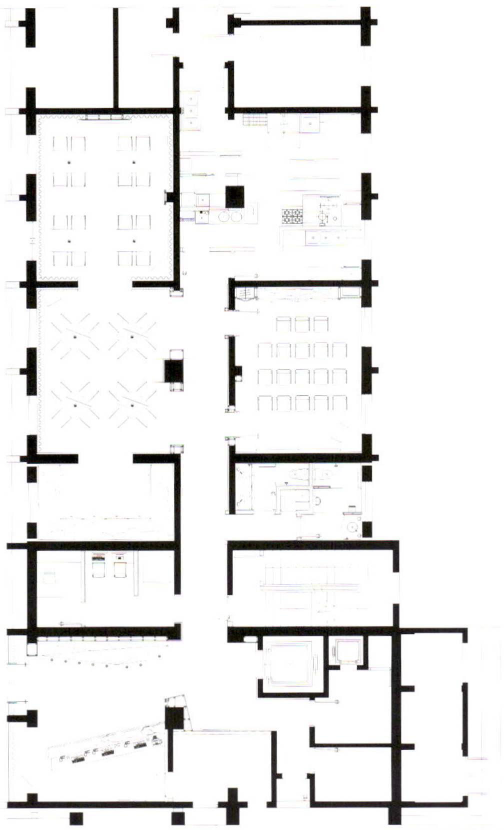

武汉市，在宽阔的紫阳路上，有一处十分低调的建筑，面积约5000m²，客房数为150间。建筑布局十分紧凑，入口处是一个狭长形的酒店大堂，面积十分有限，餐厅呈“L”形布置。整个色调为中性灰，由浅至深构成不同层次。设计的主要概念采用界面的错视感，借助不同三角平面的明度组合，形成浮雕般的楔形起伏，使人将二维平面误会成三维立体的幻觉。

There is a very humble building with a compact arrangement in the wide Ziyang Road, Wuhan. At the entrance is a narrow long hotel lobby with very limited area, which formed an “L” shape together with the restaurant. The whole color is mainly neutral gray, from light to dark which has constituted the different levels. The main design philosophy is to apply the illusionary visual sense of the interface to form a relievo-like cuniform undulation by the bright assembly of different triangular planes so that the visitors will have an illusion of mistaking the two-dimensional plane for three-dimension.

借助不同三角平面的明度组合，形成浮雕般的楔形起伏。
It forms a relievo-like cuniform undulation by the bright assembly of different triangular planes.
入口处为狭长形的酒店大堂。
At the entrance is a narrow long hotel lobby.
镜面反射下的餐厅。
The restaurant under the reflection of the mirror.
虚实交错的视觉冲击。
The visual impact in the false and realistic combination.

被折与错位 Folded and Staggered

锦江之星-银川鼓楼店

设计单位:上海泓叶室内设计咨询有限公司 / 面积:1200 m² / 坐落地点:银川 / 完工时间:2009年1月 / 摄影:叶铮

Design Unit: HYID / Area:1200 m² / Location:Yinchuan / Finish Time:Jan,2009 / Photography:Ye zheng

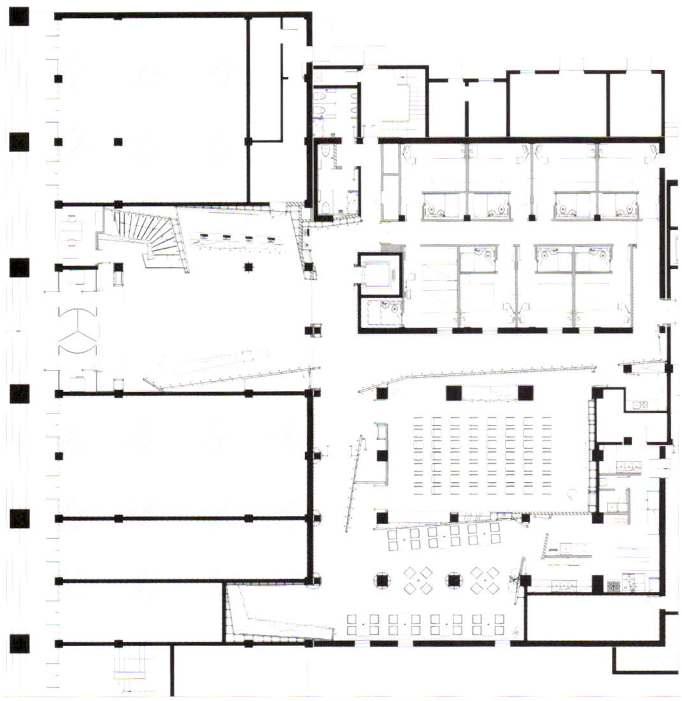

银川店的设计首先在空间流线上作出较大调整。经由大堂相对阔畅的空间后，引入一条狭长的通道，使整体空间形成纵深向的延展，构成峡谷般的形态，并由此在侧向联通餐厅区域及纵向的客房区域。设计概念主要表现为空间造型上，打破横平竖直的传统界面形态，在造型上将片特异化，形成水平和垂直双向转折。在被折的基础上，又进行撕裂错位，由此，被折与错位成为该方案的主题概念。同时辅助于材质的虚实、软硬、透堵等对比关系，使整体设计感觉更为丰富生动。不同明度的暖灰和纯黄色构成了室内的色彩概念，灯光的运用进一步凸显出空间造型的特征与魅力。

The design of Yinchuan hotel firstly has made quite big adjustment on the space streamlines. From the comparatively spacious space in the lobby, a narrow long passageway is introduced so that the entire space forms a longitudinally deep expansion in a valley-like form. It is hereby also laterally connected with the restaurant and the longitudinal guest room area. This design philosophy is mainly delivered by the space model to break through the transversely flat and vertically straight traditional interface. The models of the patches are differentiated to form a two-way turning both horizontally and vertically. On the basis of being folded, it is also torn and staggered. Thus, the being folded and staggered becomes the thematic conception of this case. Further assisted with the contrast relationships between falsehood and reality, softness and hardness as well as transparency and blocking, the whole design looks more wonderful and vivid. The variably bright warm gray and pure yellow has constituted the interior color conception while the application of light has further outstood the characteristics and charm of the space model.

不同明度的暖灰和纯黄色构成了室内的色彩概念。
The warm gray in different brightness and the pure yellow have formed the interior color conception.
材质的虚实、软硬、透堵等对比关系，使整体设计感觉更为丰富生动。
The contrasting relationship between the falsehood and reality, softness and hardness as well as the transparency and blocking has endowed the entire design with more wonderfulness and vividness.

狭长的通道使整体空间形成纵深向的延展。
The narrow and long corridor has made the entire space expanded deeply.
玻璃拉门令空间更具透视感。
The glass pull door has made the space more transparent.
打破横平竖直的传统界面形态。
Break off the traditional horizontally flat and vertically straight interface.
在被折的基础上，进行撕裂错位。
Torn and staggered on the basis of being folded.
灯光的运用进一步凸显出空间造型的特征与魅力。
The application of the light has further stressed the characteristics and charms of the space model.

舞动的截面 Dancing Section

锦江之星-郑州火车站店

设计单位:上海泓叶室内设计咨询有限公司 / 面积:780 m² / 坐落地点:郑州 / 完工时间:2009年6月 / 摄影:叶铮

Design Unit:HYID / Area:780 m² / Location:Zhengzhou / Finish Time: Jun,2009 / Photography:Ye zheng

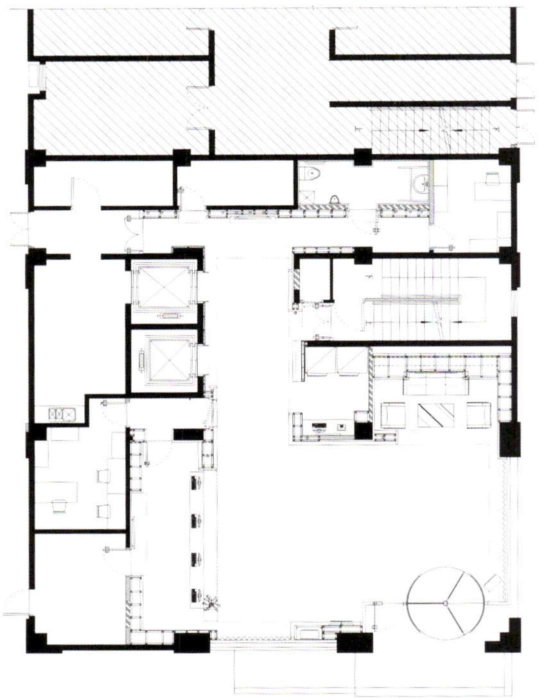

本案位处河南郑州火车站附近，建筑面积约7000 m²，客房187间。一、二层为裙房，主要功能为大堂和餐饮。室内设计的主要概念可被视为“舞动的截面”。具体而言，就是采用一片与立面垂直相切的白色起舞的流动截面，来进行空间界立面划分，并连续延展，贯通一气。同时联系功能，将蜿蜒的截面构筑成空间洞穴之形态，由此营造出中央空间与周边洞穴空间两大对比层次。这两大对比层次又分别通过色调、材质、照明这些专门手法得以进一步发展。

It is located near Zhengzhou railway station in Henan. Its first and second floors are the annex and their main function for the lobby and catering. The main philosophy of its interior design can be deemed as a "dancing section". Specifically speaking, it is to apply a white dancing flowing section perpendicular to the facade to divide the facades of the space and continuously expand integrally. Furthermore, the functions are connected together while the winding sections constitute a space cave in order to create two contrast levels between the central space and the ambient cave space. These two contrast levels are further developed through the special techniques such as the colors, materials and lighting.

大堂实景。

Lobby Scene.

白色的流动截面构成洞穴空间。

The white streaming section has constituted the cave space.

漫卷的优雅
——上海锦江之星世博村店

Pervading Graceful Rolls

设计单位:上海泓叶室内设计咨询有限公司 / 面积:9000 m² / 坐落地点:上海 / 完工时间:2009年7月 / 摄影:叶铮

Design Unit:HYID / Area:9000 m² / Location:Shanghai / Finish Time:Jul,2009 / Photography:Ye zheng

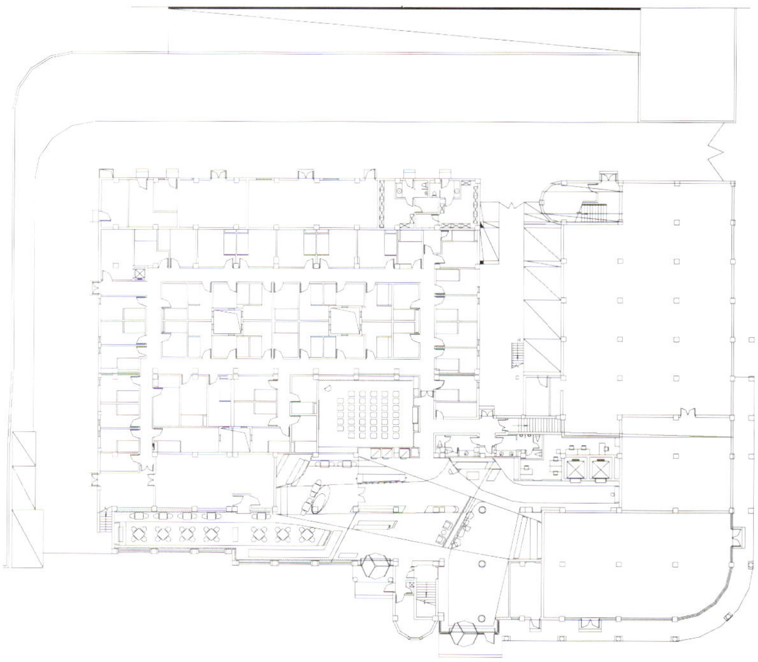

为配合上海世博的到来，位于浦东世博区旁的灵沂路与浦三路交界口，新推出了一家极具设计感的特色酒店。该店建筑面积约为9000m²，客房共计250间，外观形象十分低调，室内设计风格却极为时尚强烈。设计选取了当下最前卫的空间概念——"皮"。采用了多组轻盈连续的卷片形态，营造出单体空间之间的多意成角对比和组合。并将原建筑结构空间彻底弱化，旨在营造"一次空间"与"二次空间"的"底图"构成与黑白对比。进而使伸延流畅的卷片造形，在空间中获得十分丰富惊艳的视觉效果。在此，功能与形式、理性与激情、时尚与传统共同汇聚成最终的优雅体验。为世博，为上海，呈现出一场当代的空间观念。

Welcoming the arrival of The World Exposition Shanghai China, a newly unique hotel featured with a strong sense of design is recommended for you. The location of the hotel is in the juncture of Lingyi Road and Pusan Road near Pudong Expo Area in Shanghai. The building area of this hotel is approximately 9000m2 totaled with 250 rooms. The common exterior appearance forms a contrast with the fashion interior design The design employs a space idea "pi" which is the on the leading edge of trends. Designed to create the "base map" constituted by "one space" and the "second space" and a contrast between the black and white, multiple sets of light and continuous rolls in the design create multi-viewed angled contrasts and composition among different single spaces and completely weaken the original architectural structures. Therefore, the extended smooth roll shape gains a rich and extraordinary visual effects view in spaces. Here in this space, function and form, reason and passion, fashion and tradition together form an ultimate grace taste. This special design of the hotel shows a modern concept for Shanghai, for The World Exposition.

入口。
Entrance.
卷片弱化建筑空间结构。
Winding film weaken the structure of architectural space.

多组轻盈连续的卷片形态。
The leading edge of trends.
单体空间之间的多意成角对比和组合。
The contradistinction and combination between monomer spaces.

光影的独特魅力。
The distinct charm of shadow.
黑白对比的构成。
A contrast between the black and white.

功能与形式、理性与激情、时尚与传统共同汇聚成最终的优雅体验。
Function and form, reason and passion, fashion and tradition together form an ultimate grace taste.
伸延流畅的卷片造形，在空间中获得十分丰富惊艳的视觉效果。
The extended smooth roll shape gains a rich and extraordinary visual effects view in spaces.

在水一方大酒店扩建工程 The reconstructed project of Suzhou Lakeside Hotel

设计单位:苏州苏明装饰有限公司 / 设计:陆逊 / 参与设计:缪风 / 面积:9000 m² / 主要材料:意大利白、马皮毛的面料，不锈钢，花形马赛克 / 坐落地点:苏州 / 工程造价:5733万元 / 完工时间:2008年12月 / 摄影:贾立明

Design Unit:Suzhou Sumin Decoration Co., Ltd / Design:Lu Xun / Participatory Design:Miao Feng / Area:9000 m² / Main Materials:Italian white and horse hair fabric, stainless steel, flower-shaped mosaic / Location: Suzhou / Project Cost:RMB 57 330 000 / Finish Time:Dec,2008 / Photography:Jia Limin

楼梯过道。
Stair corridor.
用条形造型顶通过变形，产生水滴的独特韵味。
Bar shaped ceilings generate a unique style of water drops via transformation.

本案运用了极现代的设计语言，紧紧围绕“水城，花城”的理念展开联想。
本案“花”的意喻主要为宴会厅的玛瑙雕刻花形墙面，宴会厅门厅的石材雕刻花形柱，接待门厅的不锈钢花球等等，各种花形的万千变化，不同重组，演义出繁花似锦的昌盛意境。
而“水城”的设计，我们以会见厅门厅的流水墙和造型吊顶穿插在公共空间中，用条形造型顶通过变形，产生水滴的独特韵味，让人感受到大自然给人心理上的特别恩惠。使整个空间在磅礴的气势中多了些低调的成分，给人一种区别于传统奢华的内敛气度。

In this case we use a very modern design language, closely focus on the concept of "Water City and Flower City," to expand our association. In this case, the concept of "Flower" is mainly displayed on the banqueting hall agate and the carved flower-shaped walls, and a banquet hall in the lobby of stone carving flower-shaped columns, a reception hall of stainless steel flower-shaped ball, etc. We use thousands of changes in the shape of a variety of flowers and different reorganization in order to show a blossoming prosperity.
The "water city" of the design, we use the water walls and ceiling shapes interspersed in the public space of the lobby, and use deformation of the bar shape top to display the drama of unique flavor drops, to make people feeling like been gives a psychological nature special ex-bribery, which can make the whole momentum space in the Majestic to become more low-key in the composition, and compose a sense of tolerance which is different from the traditional luxury introverted.

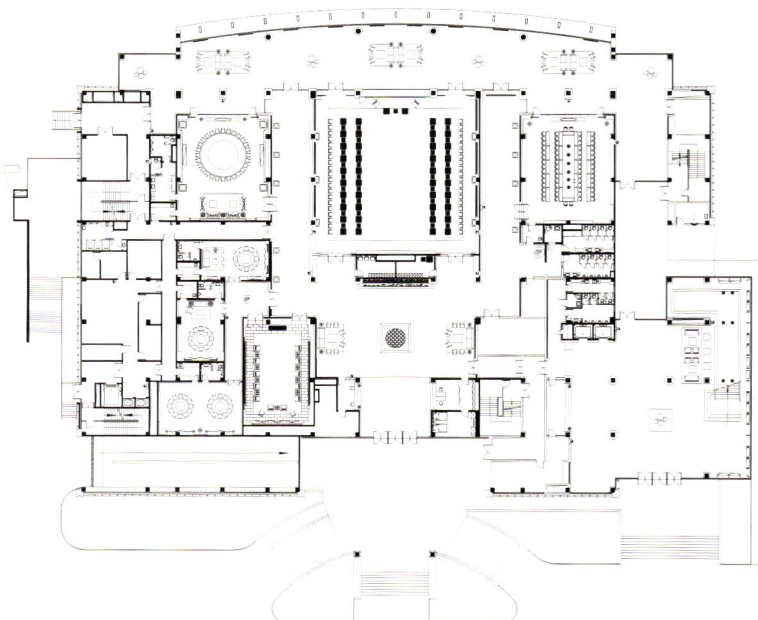

偌大的水晶灯弱化了层高太高的生硬感。
Large crystal lamps weaken the rigidity of the deep story height.
主次有分的照明体系。
Primary and subsidiary illuminating system.
公共休息区。
Public resting zone.
宴会厅门厅的石材雕刻花形柱。
Stone carved columns in the entrance hall of the banquet hall.
巨幅抽象派油画色彩跳脱。
Large size abstract painting with living colors.

大小各异、有明有暗的珠子丰富了深色墙面。
Big, small, bright and dark beads enrich the dark colored wall surface.
红色小花球静静伫立。
Standing red small ball-flower.
会议室小局部。
A small local part of the meeting room.
层层围进的大型水晶灯气势恢宏。
Large crystal lamps surrounding in layers create a grand atmosphere.

中茵皇冠假日酒店 Crowne Plaza Suzhou

设计单位:美国HPS室内设计公司 / 面积:59400 m^2 / 主要材料:木材、石材、地毯、软包、玻璃 / 坐落地点:苏州 / 工程造价:9000万元 / 摄影:潘宇峰

Design Unit:HPS Interior Design Company / Area:59400 m^2 / Main Materials:wood、stone、carpet、soft roll、glass / Location: Suzhou / Project Cost:RMB 90 000 000 / Photography:Pan Yufeng

苏州中茵皇冠假日酒店位于苏州工业园区美丽的金鸡湖畔，是一家融合了时尚邮轮元素并可满足各类商务需求的主题酒店。酒店四周风景如画，环境舒适静逸，其整体建筑设计将经典与现代风格完美融合。

402间高雅客房都独具特色，每间客房都可将园色、湖光，尽收眼底。1066 m^2的无柱宴会厅和800 m^2的宴会厅前厅气势恢宏。设施齐全的健身中心、室内、外游泳池、享誉国际的茶树水疗以及独立的游艇码头，都将为宾客留下一个完美的旅程。

Located near the Jinji Lake in the Suzhou Industrial Park, Crwone Plaza Suzhou is a theme hotel that combines fashionable elements of a passenger liner, and meets all business requirements. The hotel is featured by a picturesque landscape nearby, and a comfortable and peaceful environment. The overall building design perfectly combines the classic style and the modern one.

All 402 guest rooms show unique characteristics, where you can have a panoramic view of the landscape. Covering an area of 1066 and 800 square meters respectively, the pillar-free banqueting hall and its anteroom are indeed full of power and grandeur. You will have a perfect holiday by enjoying the following facilities: the fitness center with all necessary facilities, indoor and outdoor swimming pools, world-renowned tea tree oil therapy and independent yacht wharfs.

弧形天棚使用木条拼贴，其间穿插钢管。
Cambered ceiling with wooden bars and crossing steel pipe.
形似帆船脊背的天棚气势非凡。
Splendid ceiling likes the back of a sailing ship.

大厅顶部设计煞费苦心。

Pains-taking design for the hall ceiling.

简单中又因色彩而显得高贵的客房。
Distinguished guest's room based on simpleness and colors.
水中游来游去的红鲤鱼有着美好的寓意。
Swimming paradise fish of China is a symbol of goodliness.
在轴线的院落序列上运用隔障的手法，呈现虚实，营造层次。
Courtyard designed with barriers on the axis is represented in false and true separations.

金海华丽嘉酒店 King Grace Hotel

设计单位:郭轲蔚酒店设计事务所 / 设计:郭轲蔚 / 面积:25000 m² / 主要材料:木材、石材、墙纸、地毯 / 坐落地点:常熟 / 工程造价:2.5亿元 / 完工时间:2009年1月

Design Unit:Guo Keiwei Hotel Design Office / Design:Guo Kewei / Area:25000 m² / Main Materials: wood、stone、wallpaper 、carpet / Location: Changshu / Project Cost:RMB 250 000 000 / Finish Time: Jan,2009

在欧罗巴的苍穹下，孕育了拥有独特性格的众生。欧式家具的精髓，更是欧罗巴文化的缩影，白色成品木制线框完美演义了西方古典主义的体贴入微——细节决定一切。本案定位精致会所型酒店，于繁复间寻求低调奢华的平衡，充分利用了色彩和图腾的变幻，营造出饱满浓郁的贵族气息。硬装上，大胆采用夸张的线条，精雕细琢而不乏细腻之感。质地柔和的墙纸与变化多端的地面交相辉映,强调协调统一的高品位。

和谐、高雅是欧罗巴古典风格的代名词。在欧式风格常见的金、褐、暗红等主色调中，少量糅合白色，使色彩更明快大方，使整个空间给人以开放、宽容的非凡气度，让人丝毫不显局促。

The hotel gives birth to the unique character of all living creatures under the starry sky of Europa. The furniture in the hotel possesses the essence of European furniture and especially become the epitome of the European culture. Its White finished goods and wooden frame fully represent the consideration of the Western Classic. This cases position on the exquisite club hotels and discover the balance of the complication and a low-key luxury. The design fully used the changes of the color and totem so that create a rich flavor of the noble. In terms of hard-mounted decoration, the design use of exaggerated lines to crafted carefully and no lace of exquisite. The soft texture wallpaper adds radiance and beauty to the diverse floor and this formed a high-grade stressing harmonization.

Harmonious and elegant is a synonym for Europa classical style. A small amount of blends white with common gold, brown, dark red and other main colors in the European-style makes a bright and generous style. The whole space brings people an magnanimous feeling of open and tolerant and also a feeling of relax.

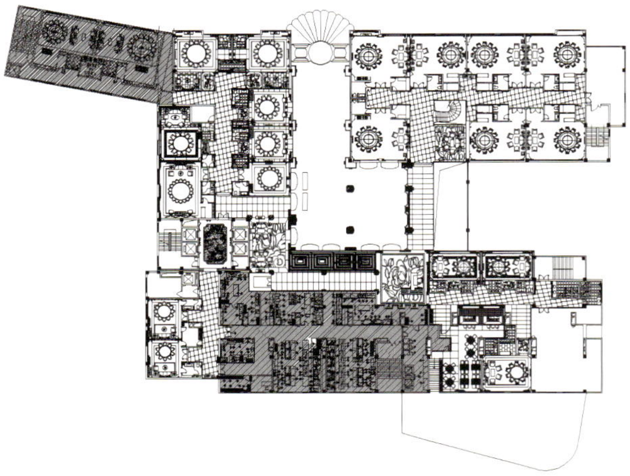

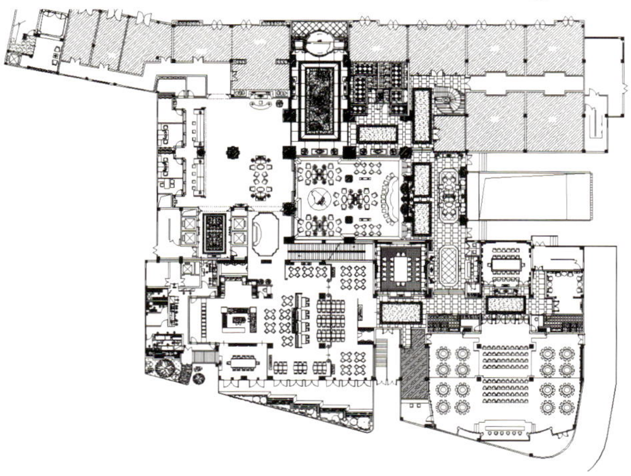

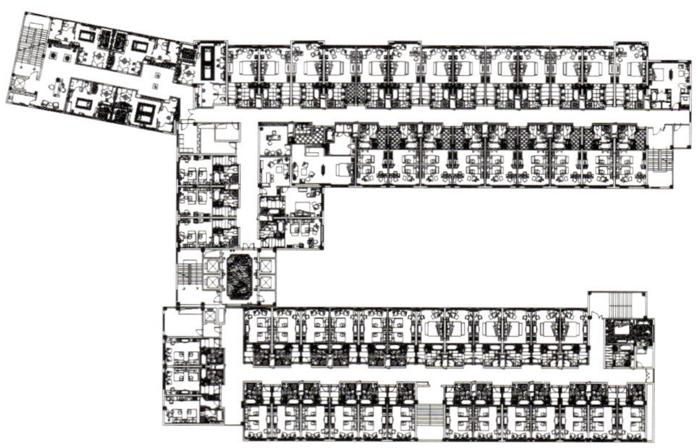

繁复间寻求低调奢华的平衡。

Seek for the balance between the low-profile and luxury among the heaviness and complicatedness.

白色成品木制线框完美演绎西方古典主义的体贴入微。

The white finished wooden frames are perfectly depicting the every possible consideration of the western classism.

大胆采用夸张的线条。

The exaggerated lines are boldly applied.

明快、大方的走道。

The bright and natural corridor.

利用色彩和图腾的变幻，营造饱满浓郁的贵族气息。

The irregular change of t the colors and totems are applied to create a full and strong noble atmosphere.

精雕细琢而不乏细腻之感。

The sense of exquisite and delicate carving, without lack of minute description.

质地柔和的墙纸与变化多端的地面交相辉映，强调协调统一的高品位。

The gentle wall papers and changeable floor add radiance and beauty to each other perfectly with an emphasis on the harmonious and unified superior taste.

花园城销售中心 Garden City Sales Centre

设计:琚宾 / 参与设计:许志强、张海军 / 面积:400 m^2 / 主要材料:水晶吊线、水泥自流平、白色乳胶漆 / 坐落地点:深圳 / 工程造价:1 00万元 / 完工时间:2009年8月 / 摄影:陈勇

Design:Ju bin / Participatory Design:Xu Zhiqiang 、Zhang Haijun / Area:400 m^2 / Main Materials:crystal wire、TFSF、white emulsion paint / Project Cost:RMB 1 000 000 / Finish Time:Aug,2009 / Photography: Chen yong

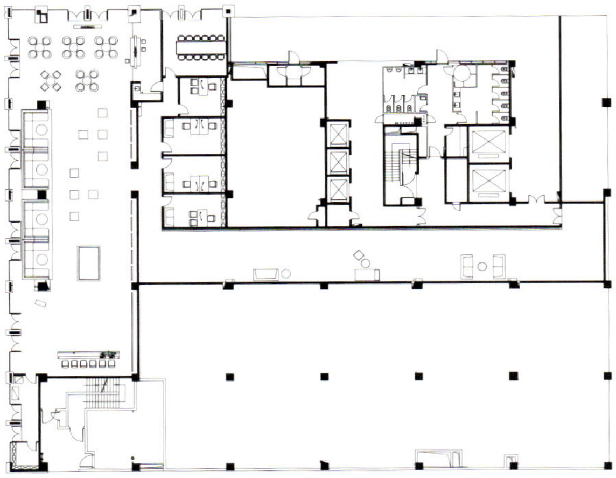

光给予我们无序变化；雨给予我们韵律之美；光和雨是我们灵感的来源。简洁的水泥自流平地面；素雅的白色乳胶漆墙面；加上一根一根挂线的过程；是艺术；是行为艺术；是装置艺术，渗透在我们售楼中心的每一个角落。每个人，每一物，都是它的一部分。是艺术与功能的结合，是装置艺术在室内空间的运用。

Light changes to give us the disorder; Rain gives us the beauty of rhythm; Light rain is our source of inspiration. Simple self-leveling cement floor; Elegance of a white wall paint; Add A process of a hung up; Art; Is performance art; Is installation art, Infiltration in our sales centers in every corner. Everyone, Each object, Are part of it. Is a combination of art and function The installation of art in the use of interior space.

纯白世界中一抹红色特别亮眼。
A little red is especially outstanding among a pure white world.
微型展示台仿佛被白雪包围。
The mini exhibition stand looks like being enclosed by white snow.
简洁的水泥自流平地面搭配素雅的白色乳胶漆墙面。
The simple cement self-leveling ground is arranged with the simple but elegant white latex paint wall.

一根一根挂线的过程也是一门艺术。
The process of hanging line by line is also an art.
高低不一的垂线丰富了空间。
The pendant lines in different heights enrich the space.

加上自然光的引入，空间如诗如画。
Together with the introduction of natural sunlight, the space looks poetic and picturesque.
幽蓝的光线朦胧了空间。
The blue light has dimmed the space.
精致的样板模型。
The delicate prototype.

龙湖睿城销售中心 Longfou Wisdom City

设计单位:深圳市朗联设计顾问有限公司 / 设计:秦岳明 / 参与设计:唐浩 / 面积:800 m² / 主要材料:灰木纹、黑白根、柚木、防腐木、夹金属玻璃、特效腐蚀玻璃 / 坐落地点:重庆 / 工程造价:260万元 / 完工时间:2008年6月 / 摄影:井旭峰

Design Unit:Shen Zhen Rongor Design & Consultant CO.,LTD / Design:Ronger Kane / Participatory Design: Tang hao / Area:800 m² / Main Materials:wood、black marquina、teak、syp、glass / Location:Chongqing / Project Cost:RMB 2 600 000 / Finish Time:Jun,2008 / Photography:Jing Xufeng

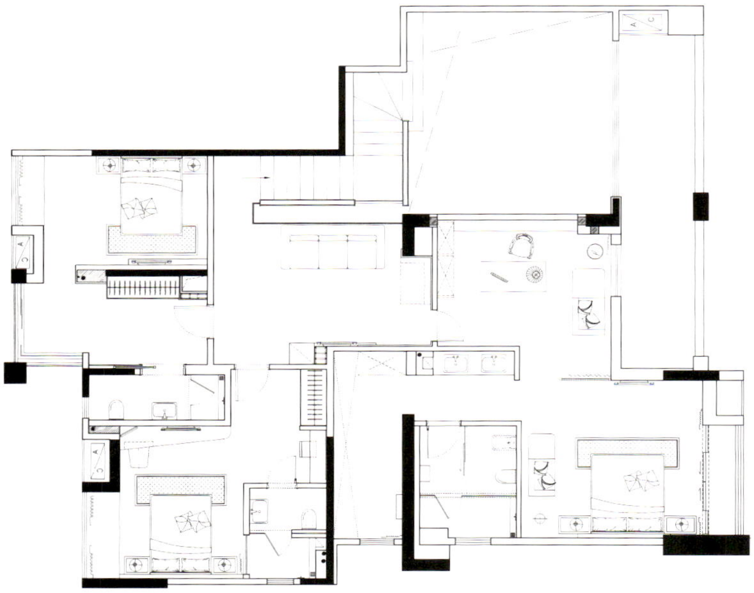

睿城项目是一个纯现代的建筑空间，业主期待能做到对“传统”和“东方”的回归。面对传统，我们不希望只是简单形式与符号的堆积，脱离现代人文环境和生活的传统，就像古老的弧线瓦，虽有着优美的线条，却已不再符合现代工业的生产方式。

设计者在空间的中间及后部各增设两个彼此呼应的庭院，所有的功能均围绕空间的“中心院落”展开。人工发光天棚模拟洒满庭院的阳光，渗透各处留下斑驳的光影。墙面尺度不一的木格则来源于对传统青瓦屋顶的影像记忆，收集着断续的光阴故事。演变自“Wisdom City”的字母符号宛如古老的图腾布满中庭的主要墙面，用现代的手法强化项目的标识性。

轻盈与透明消解了传统空间中的“不能承受之重”，后部小型室内庭院配合中心庭院形成生动的院落序列，强化空间的层次感与转承关系，并与室外园林相呼应，延续了室内外空间的交流与对话。现代与传统在这里没有界限，如今的现代在不久的将来会是另一种传统。

Longfou Wisdom City, which is a purely modern architectural space, now has been expecting to be a regression of "tradition" and "Orient" by the owners. Due to the "Tradition", we hope it is not just accumulation of simple form and sign and separated from the modern human surroundings and lives, for example, we all love the ancient pitch arc, which has graceful lines but is not suitable for the productive method of modern industry.

Our designer also put two courtyards which correspondent to each other in the middle and back of the space. And all functions are expanded around the "Central country yard" of the space. The artificial luminous ceiling is a perfect simulation of the sunshine sprinkling the whole yard, which permeates each place and leaves parti-colored shadow. And the wooden lattice with different sizes on the wall root from the image memory of traditional roof with grey tile, which collects the stories of intermittent time. The letter signs of developed from Wisdom City are like the ancient totem, spreading the main walls of the middle yard. Modern skills are taken to strengthen the identification of the project strongly.

"Lightness" and "Transparency" eliminate the "unsustainable weight" of the traditional space. And the mini-type indoor yard in the back coordinative with the central yard forms the vivid yard sequence, which strengthens the hierarchy feel and passing relationship of the spaces and is correspondent with the outdoor gardens, and prolongs the communication and conversation between the indoor and outdoor space. There is no limit for modernity and tradition here and we believe the current modernity must be another tradition in the later future.

玲珑剔透的细部。

The exquisite and bright details.

顶部木格的围合控制了光影，也控制了区域感的形成。

The enclosure of the wood panes on the top have both controlled the light and shadow as well as the formation of regional sense.

隔断形成的虚拟空间。

The dummy space formed by the partitions.

视线的穿透与断隔。
The penetration and isolation of the sight lines.
由wisdom city 演变而来的字母符号，极具识别意义。
The alphabetic symbols evolved from wisdom city have extreme distinguishable meaning.
休息区一角。
One corner of the rest area.
木格纹样的延续使用。
The continued application of wood grained patterns.

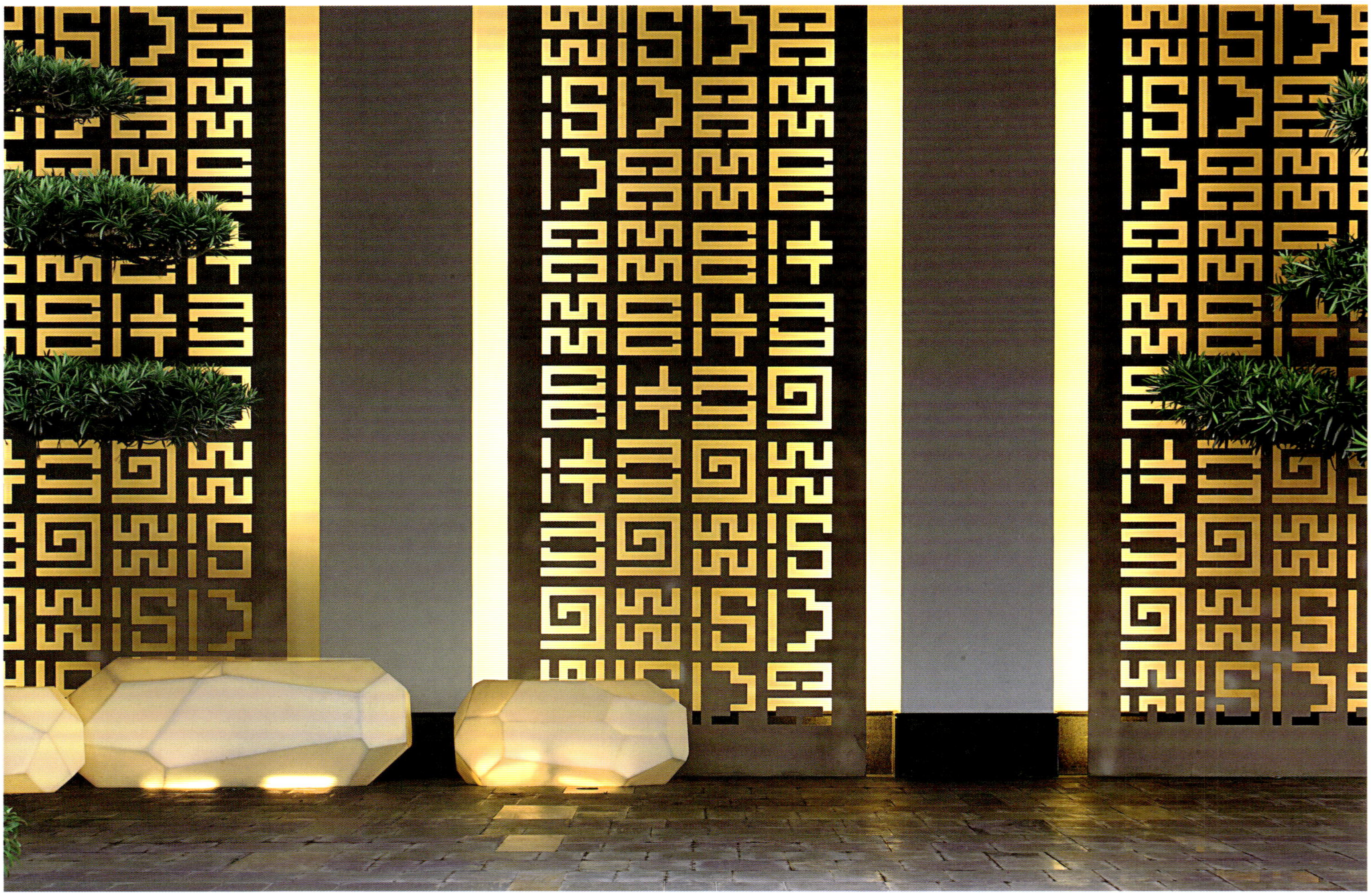

奥林匹克花园冠军城 Champion City of Olympic Garden

设计单位:广州力思工程设计有限公司 / 设计:骆志兴 / 参与设计:关家纹、郑敏仪 / 面积:280 m^2 / 主要材料:地板灰木纹石、黑晶石、灰镜、不锈钢 / 坐落地点:番禺 / 工程造价:140万元 / 完工时间:2009年1月 / 摄影:林力勤

Design Unit:LESS / Design:Luo Zhixing / Participatory Design:Guan Jiawen、Zheng Minyi / Area:280 m^2 / Main Materials:stone、black spinel、neutral density filter、stainless steel / Location:Fanyu / Project Cost: RMB 1 400 000 / Finish Time:Jan,2009 / Photography:Lin Liqin

设计延续现代时尚的思路，运用简单、沉稳、灵动的搭配。由于室内空间狭长，特意做了一些指引。地板灰木纹石中镶嵌部分黑晶石，黑晶石组成的线路有序地从门口经过模型台，辗转至洽谈区，再通过入户台阶直入大堂，指引人们有序地进行参观。

天花吊顶的设计与地面相映成趣。吊顶的路线紧紧跟随着黑丝带，在黑晶石所形成的平面图形中变化，在空间里不断地做着升降的运动。这样原本二维平面的图形便生成了立体三维的物体。造型上黑色镜面材质和凹入墙身的黑色镜面不锈钢的体量组合加深空间的层次感，令空间大胆跳跃，具有艺术意味。

The design has continued the modern and fashionable philosophy with the application of the simple, steady and bright arrangement. Some directions have been made due to its long and narrow interior space. Some black spars are embedded in the gray wood textured stone on the ground to form into a route in a sequence from the entrance to the model table, the discussion area and then through the door step to the lobby so that people are guided to visit and get to know in an orderly way.

The ceiling is designed to form an interesting contrast with the floor. The ceiling route has closely followed the black ribbon and made changes in the plan layout formed by the black spars, which are constantly making ascending and descending activities. In this case, the original two-dimensional figures become the solid three-dimensional objects. In regard to the model, the combination of the black mirror material and the practical stainless of the black mirror concaved into the wall has further deepened the level sense of the space, which makes space bold and jumping with an artistic atmosphere.

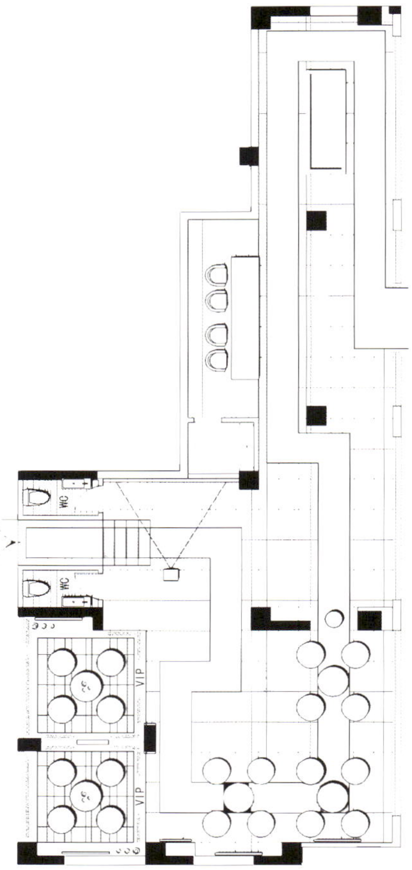

强烈的导向性表达。

The strong directional expression.

线帘形成的呼应感加强了展台的重要性。

The echo among the line curtains has strengthened the importance of exhibition stand.

由材质区分过道，突出立柱与地面的关联。

The corridor is separated from the material zone and stresses the relationship between the columns and ground.

镜面效果的吊顶，空间得到延伸。

The mirror ceiling has extended the space.

秋实e景售楼处 Sales Pavilion for Qiu Shi e Jing

设计单位:长春陈旭东设计事务所 / 设计:陈旭东 / 参与设计:刘秀梅 / 面积:650 m² / 主要材料:玻璃、白地砖、木材 / 坐落地点:长春 / 工程造价:150万元 / 完工时间:2009年1月 / 摄影:朴磊

Design Unit:Changchun Chen Xudong Design Office / Design:Chen Xudong / Participatory Design:Liu Xiume / Area:650 m² / Main Materials:Glass、white floor tile、wood / Project Cost:RMB 1 500 000 / Finish Time:Jan,2009 / Photography:Piao lei

秋实e景的一棵树。
A tree in Qiu Shi e Jing.
走道全景。
The overall scenery of the corridor.

设计师以"e"的信息母语为起点，在计算机语言形式中抽取线性结构，通过对空间的认识和理解，把功能场所分为明、暗两个空间区段。明为深度洽谈区，暗为体验观赏区，中间由总服务台连接。就线形的灯带划分而言，可以看到两个方向上的空间变化，一个是顺向的，一个是横向的，两个运动特性在同一个空间中产生对撞，使进入者第一感受就是时间的流失以及生命感的自我重塑。空间中材料的运用与表达始终贯穿于设计风格之中，条形的木质饰面板由线性转换到面，再由面与条形灯带形成图底关系，一宽一窄，一暗一明从墙壁过渡到顶棚，使材料的肌理在灯光的映衬下形成空间界面的自由转换，增强了空间意象由触觉感向视觉感转换的过程。

The designer has taken the mother language of information – "e" as his starting point and has abreacted the linear structure from the computer language form to divide the function area into the bright and dark space sections through the recognition and understanding of the space – the bright is for the further discussion area while the dark is for the enjoyment and admiration area, which are connected by general service desk in the middle. As for the division by the linear light ribbons the space change can bee seen in two directions, one is forward and the other is transversal. The two athletic characteristics are collided with each other in the same space so that the entrant can first feel the passing of the time and the self–remolding of life awareness. The application and demonstration of the materials in the space have run through the design style. The strip wooden veneers are transferred from the linearity to the surface in form. Then the surface and strip light ribbons formed a relationship of figures and background, which transit from the wall to the ceiling in a way of one wide and one narrow as well as one dark and one bright, so that the textures of the materials form a free exchange of the space interfaces in the contrast of the light and the conversing process from tactile sense to visual sense of the space image is strengthened.

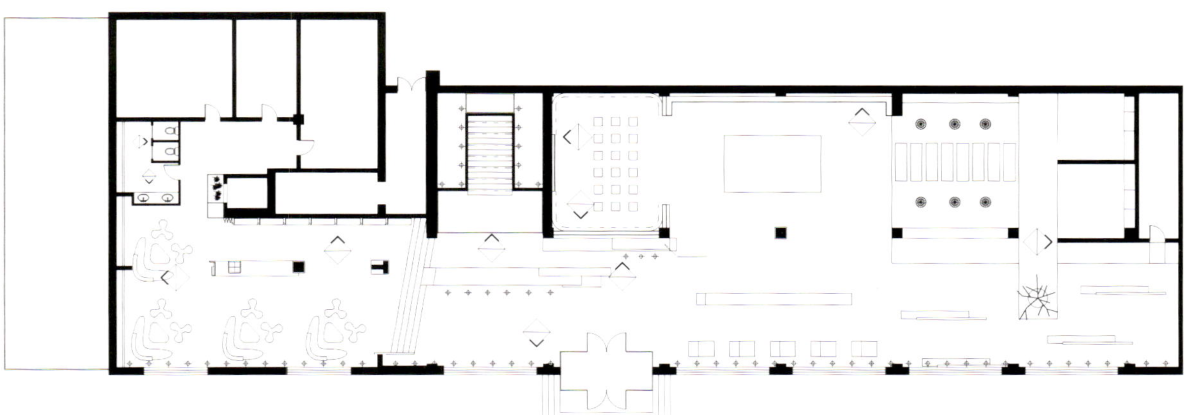

进入的过程就是“渡过”的过程。

The access process is just the “transitional” process.

一宽一窄，一暗一明从墙壁过渡到顶棚。

One width by one narrowness as well as one darkness by one brightness have transited from the wall to the ceiling.

光带的运用起到很好的视线引导作用。

The application of the light ribbons has taken a good visual guiding effect.

红色的椅子在纯白的空间中分外抢眼。
The red chairs become extremely outstanding among the pure white space.
线形的灯带划分，一个顺向，一个横向。
The division of linear light ribbons, one horizontal and one vertical.
时光隧道般的空间。
The time tunnel-like space.

锦园销售中心 Jin Yuan

设计单位:珠海市空间印象装饰设计工程有限公司 / 设计:王刚 / 参与设计:李江威 / 面积:800 m^2 / 主要材料:铁刀木、灰木纹大理石、染色镜钢、镀木膜玻璃、金箔、墙纸 / 坐落地点:珠海 / 工程造价:120万元 / 完工时间:2009年5月

Design Unit:SIDD / Design:Wang gang / Participatory Design:Li Jiangwei / Area:800 m^2 / Main Materials: cassia siamea、marble、steel mirror、glass、gold foil、wallpaper / Project Cost:RMB 1 200 000 / Finish Time:May,2009

本案设计以自然、奢华为主题。平面功能布局上，把主体销售展示和商业洽谈区分开。洽谈区抬高两级用镀膜玻璃护栏隔离，保持洽谈区的私密性和优越感。吧台作为洽谈区的辅助功能，开阔的视角兼顾了洽谈区与展示区的客户。

大量使用透明玻璃代替原来的外墙隔断，以灯光水景过渡室内外空间，同时以大量斧劈石修饰外表面，仅以顶部超大尺寸的翻板式广告幕点醒楼盘名称。室外光线的引入，配合室内点光源来强调沙盘模型，使整体气氛更加和谐。

大面积灰木纹石材表达“自然奢华”主题。沙盘的雕花玻璃基座和金箔装饰线条衬托出华丽的气息，仿木墙面柔和视线，尽量让人工气息变得清淡。洗手间设计在隔断内另有玄机：一个静止的雕塑墙面“枯木”盘亘，一个潺潺水幕的动态场景“鸟儿”栖息，让男女客人各有惊喜。

This design is subject is natural, luxurious and magnificent. In regard to its function layout, the main sales and exhibition area are separated from the business discussion area. The business discussion area is uplifted with two steps separated by the coated glass railing to keep the intimacy and superiority complex of the discussion area. The bar counter functions as a support to the discussion area and its wide visual angle has both considered the customers in the discussion area and exhibition area.

In this design, the transparent glass has been largely used to replace the original exterior partition wall and the light and water scene is applied to transit between the interior and exterior spaces. The introduction of the exterior rays is combined with the interior dot light sources to emphasize the sand table and make the whole atmosphere more harmonious.

The large area of gray wood textured stone materials has expressed a subject of “natural, luxurious and magnificent”; the cut glass base of the sand table and the golden foil molding have set off the magnificent atmosphere; the design of the toilet inside the partition wall has another mysterious principle: one still sculpture wall is wound by “withered woods” while a dynamic scene with the ripply water curtain has “birds” rested on it, which have pleasantly surprised both male and female customers.

沙盘局部。

Partial sand table.

接待台稳重、内敛。

The reception desk is steady and restrained.

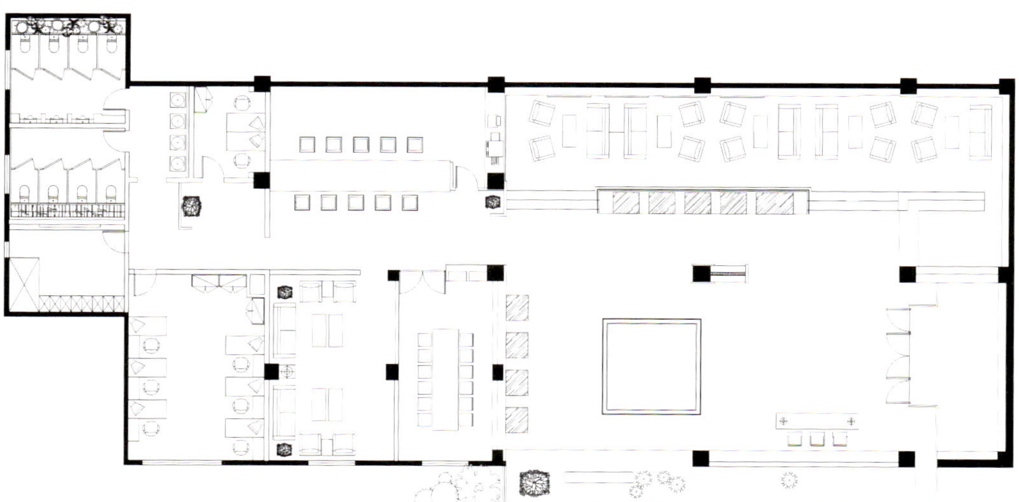

中心沙盘。
Central sand table.
大面积灰木纹石材表达“自然奢华”主题。
The large area of grey wood grained stone materials expresses the theme of “naturalness and luxury”.
洗手间的台盆古朴、自然。
The washbasin in the wash room is simple, unsophisticated and natural.
客服区顶部的雕花造型与展台玻璃纹样一致。
The carving model on the top of the customer service area is same to the glass pattern in the exhibition stand.

发现之旅售楼处 A Journey of Discovery

设计单位:达观国际建筑室内设计 / 设计:凌子达 / 面积:570 m² / 主要材料:柚木、老榆木、木丝水泥板、中国黑大理石、西班牙大理石、马赛克 / 坐落地点:厦门 / 完工时间:2008年9月 / 摄影:周跃东

Design Unit: KRIS LIN INTERIOR DESIGN / Design:Kris Lin / Area:570 m² / Main Materials:teak、elm、wood wool cement board、black marble、mosaic / Location: Xiamen / Finish Time:Sep,2008 / Photography:Zhou Yuedong

原建筑空间有良好的采光，充足的阳光透过木质百叶的间隙洒到地面，为木丝水泥板加些线条的光影，线与面在这里发生交错。功能上作为划分空间的瀑布流水衬托在深色石材上的同时，水的动态相对亦营造出空间的静谧感。

在材料运用上，原质天然素材凸显山水相映的特点。狭长的平面上，充分运用方与圆的语汇，在动线行进中有一份发现的惊喜。接待台处，方形水池与洞窗、圆形图案吊顶的结合是天圆地方的宇宙缩影。

The original building space has good day lighting. The sufficient sun light is spraying on the ground through the gaps of wood blinds to add some light and shadows of the lines to the wood-lined cement floor, where the lines and surface are interwoven together. The waterfall and running water which functions to divide the space has been in contrast with the dark stone materials while comparatively, the dynamics of the water also has created a sense of quiet space.

In the application of the materials, the originally natural elements have prominently reflected the characteristic as the mutual setoff between water and hill. In this narrow and long plan, the vocabulary of square and roundness has been sufficiently applied and there is a surprise and joy of discovery during the action of the active lines. In the reception area, the combination of the square pond and window hole as well as the ceiling in round pattern is a miniature of the universe which has a round sky and a square ground.

大面积落地玻璃引入充沛的自然光线。

The large area of floor glass introduces the abundant natural sunlight.

线与面在这里发生交错。

The lines and surfaces are strewn at random here.

原质天然素材凸显山水相映的特点。

The original natural materials prominently show the feature of mutual setoff between the mountain and water.

弧形的休息区。

The arc rest area.

在动线行进中有一份发现的惊喜。

There is a pleasure of discovery in the advance of active lines.

万科金澳国际展示中心 Vanke Exhibition Center

设计单位:维思平建筑设计 / 设计:陈凌、吴钢、张瑛 / 参与设计:白云祥、张建锋 / 面积:1098 m² / 主要材料:钢、白色烤漆玻璃、橡木、微晶石、人造石、白色乳胶漆 / 坐落地点:天津 / 工程造价:1000万元 / 完工时间:2008年10月 / 摄影:舒赫

Design Unit:WSP Architects / Design:Chen ling、Wu gang、Zhang ying / Participatory Design:Bai Yunxiang、Zhang Jianfeng / Area:1098 m² / Main Materials:steel、storing varhish glass series、oak、microlite、artificial stone、emulsion paint / Location:Tianjin / Project Cost:RMB10 000 000 / Finish Time: Oct,2008 / Photography:Shu he

本案是钢结构建筑体系系列项目之一，该体系的设计目标是在短期快速建设要求下，既可满足不同客户的功能需求，又能有效降低对生态环境的破坏。

该建筑要求能短期内建成，经使用后改变使用功能或者直接拆除。通常情况下，针对此类要求设计的建筑往往在建设、改造和拆建的过程中容易对原环境产生破坏，且建筑材料较少循环利用，造成资源浪费。设计师综合快速建造、拆建便利、形式灵活、环境保护、造价控制等目标后，应用钢结构建筑体系达到快速环保的建造效果，建筑室内也实行了一体化设计。

This building has been required to be built up in a short time and its functions are changed after construction. Normally, the buildings designed with such requirements will always damage the existing environment during the construction, modification, removal and reconstruction and the few construction materials are recycled for use so that the resources are wasted. After the integration of the targets including quick construction, convenient removal and reconstruction, flexible methods, environmental protection and cost control etc., the designer has applied the steel structure system to achieve the quick and environmental protecting construction effects. The interior of the building is also designed in an integrative way.

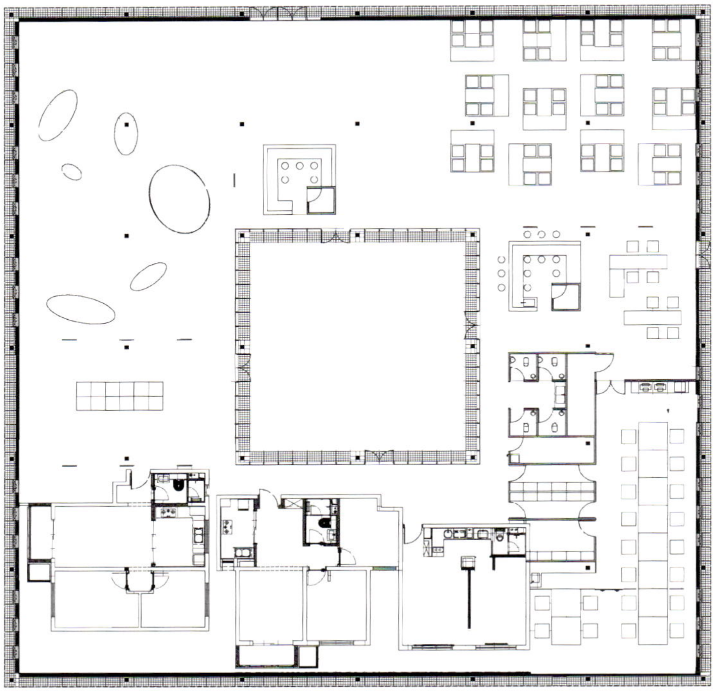

过道。

Corridor.

顶部构造梁本身产生的肌理感与墙面产生对比。

The texture generated by the beam on the top structure forms a contrast with the wall.

展厅。
Exhibition Hall.
服务台。
Service Desk.
白色立墙次第的错列加强了空间的层次。
The staggered sequence of the white vertical walls has strengthened the space levels.
外观。
Visual Appearance.

中航城 CATIC City

设计单位:深圳市派尚环境艺术设计有限公司 / 设计:李益中、范宜华 / 面积:570 m^2 / 主要材料:皮革、木地板、不锈钢、黑色地毯、白洞石、黑镜 / 坐落地点:赣州 / 工程造价:90万元 / 完工时间:2008年12月 / 摄影:吴永长

Design Unit: panshine.,ltd. / Design:Li Yizhong、Fan Yihua / Area:570 m^2 / Main Materials:leather、wood floor、stainless steel、black carpet、white cavity-stone、dark mirror / Location:Ganzhou / Project Cost: RMB 900 000 / Finish Time:Dec,2008 / Photography:Wu Yongchang

利用原建筑无序的柱体构成，通过斜拉式的大界面来达到空间的统一，简洁、大气的界面附之于斜式感，在很好地解决销售空间营销路线的条件下，让空间富有极强的透视感与设计感，界面的大尺寸与整洁构成宏大的气势，让人印象深刻，更好地体现楼盘气质。

The unification of the space is realized with application of the disorderly column constitution in the original building and through the cable-stayed large interface. The simple and grand interface is endowed with the inclining sense. The space is enriched with the extremely strong transparency and design sense in the condition of the better settlement of the marketing physiology for the space sales. The large size and orderliness of the interface has formed a grand atmosphere, which is very impressive and better reflects the quality of this estate.

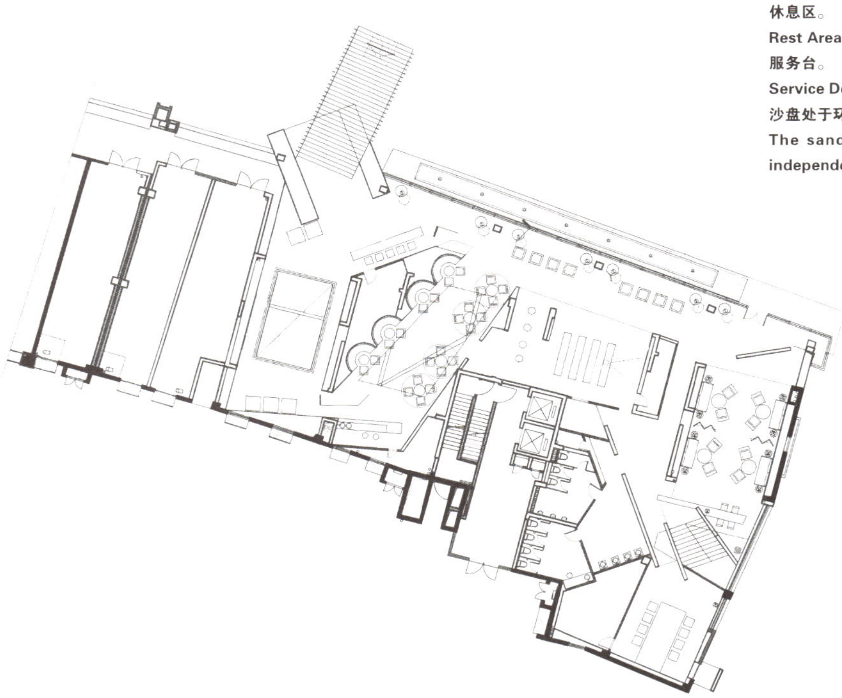

休息区。

Rest Area.

服务台。

Service Desk.

沙盘处于环形动线上，同时区域具有相对独立性。

The sand table is located on the circular active lines and the area has the comparative independence.

斜向镜面使窄小的过道不再逼仄，反而增添了趣味性。

The inclined mirror makes the small corridor not narrow any more, and increases its interest instead.

过道。

Corridor.

地面材料铺贴方向的变化。

The change of the paving directions of the floor materials.

保利·麓谷林语销售中心

Marketing Center of Poly Ligulinyu

设计单位:于强室内设计师事务所 / 设计:林敏玲 / 面积:1400 m² / 主要材料:白色大理石、木材 / 坐落地点:长沙 / 工程造价:600万元 / 完工时间:2009年6月

Design Unit:YuQiang & Partners Interior Design / Design:Lin Minling / Area:1400 m² / Main Materials: white marble、wood / Location:Changsha / Project Cost:RMB 6 000 000 / Finish Time:Jun,2009

室内的设计并未在原有建筑方案中过多地改动，只在T型的两个空间的交叉部分，即建筑入口处顶部，用浅色木片以模数切割出一片片硕大的叶片，上下分若干高差进行组合，织成网状，强调出入口的同时增加两个同等高空间的层次。叶薮时密时疏，灯光模拟高挂的阳光从间隙中穿入，结合墙上浮刻状的鸟儿，轻柔素雅的白色基调，目的意在以现实的材料拟造一种似有若无、置身其中会不觉低语的氛围。

The indoor design remains basically the same to the original building scheme, except that in the T-shaped cross section of the two spaces, namely, at the top of the entrance of the building, light wooden plates are used to cut into large leaves, which are assembled vertically according to elevation differences before being knit into a net. In this way, two spaces with equal heights are formed in addition to emphasis on passageways. The light penetrates the gap of dense or scattered leaves by imitating the sunlight, with the birds scribed on the wall to create a gentle, simple but elegant white background. The purpose is to create a visional atmosphere with real materials, where you cannot help whispering.

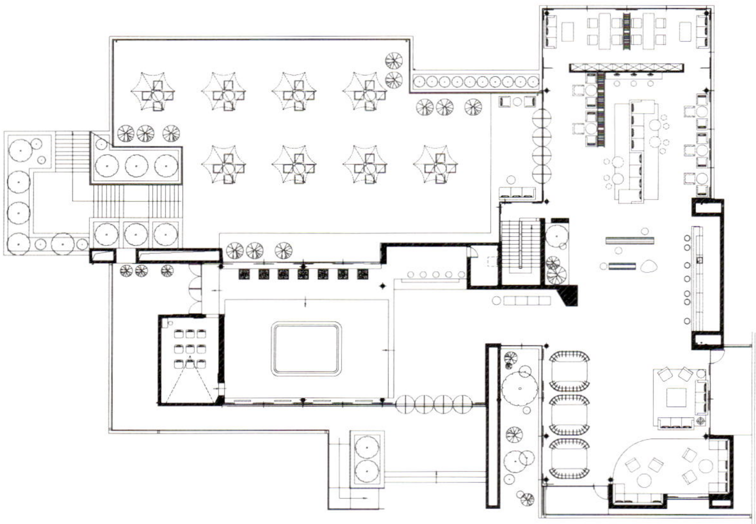

精致的顶部。
Elaborative top.
楼梯拐角。
Stair corner.
来访接待区。
Visitor reception zone.

保利地产
保利·麓谷林语
林语境界 欢享人生

叶薮时密时疏，灯光模拟高挂的阳光从间隙中穿入。

Dense or nondense leaves with sunshine-like lamp light penetrating via the gaps.

叶片上下分若干高差进行组合，织成网状。

Network combination in depth differences.

客户洽谈区。

Client negotiation zone.

室内的设计并未在原有建筑方案中过多地改动。

Indoor design is not modified too much in the original building plan.

浮刻状的鸟儿，轻柔素雅的白色基调。

Carved birds in a gentle and elegant white fundament.

中海-国际社区销售中心 Zhonghai-International community sale center

设计单位:IADC涞澳设计有限公司 / 设计:张成喆 / 面积:2000 m² / 主要材料:白云石、亚克力、铝、不锈钢、玻璃、柚木 / 坐落地点:苏州 / 完工时间:2008年7月

Design Unit:IADC DESIGNERS LTD. / Design:Elio Zhang / Area:2000 m² / Main Materials: dolomite、ACRY、aluminum、stainless steel、glass、teak / Location:Suzhou / Finish Time:Jul,2008

平面规划中，以非对称的手法，营造趣味性的空间变化。直线和曲线相结合，使空间富于张力。透明材质的使用，结合反光物料，视觉上轻盈、通透，并富于现代感。影音室的圆弧形外观，由不规则的孔状构造组成，发出淡蓝色光线，成为空间的视觉焦点。墙面的块面构成和光线的变化，犹如浮雕的效果。

设计师在本案中试图营造富戏剧性的直观感受，并通过“几何结构”演绎了这一体验空间。“存在、理想”成为此空间设计的主题之一。

In the floor planning, using the asymmetric method to make the interesting space change.Combination of straight and curve line make the space extending.The transparent materials utilization and combination with the reflective materials; it locks light and transparent as well as be endowed with modernization. The convex appearance of the AV Room are made up with the irregular holes structures, giving out the light blue light, become the visual focus of the space. The module cuted of the wall and the changing lignt have the effect of relief.

The Designer attempt to make a directly reception of dramtic in this case, and perform the experienced space through geometry. The existence and ideality become one of the subjects in the space design.

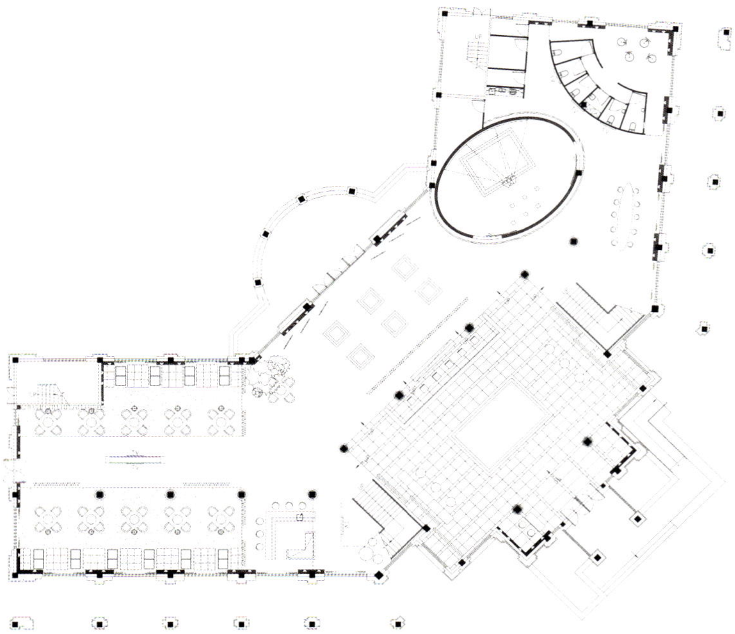

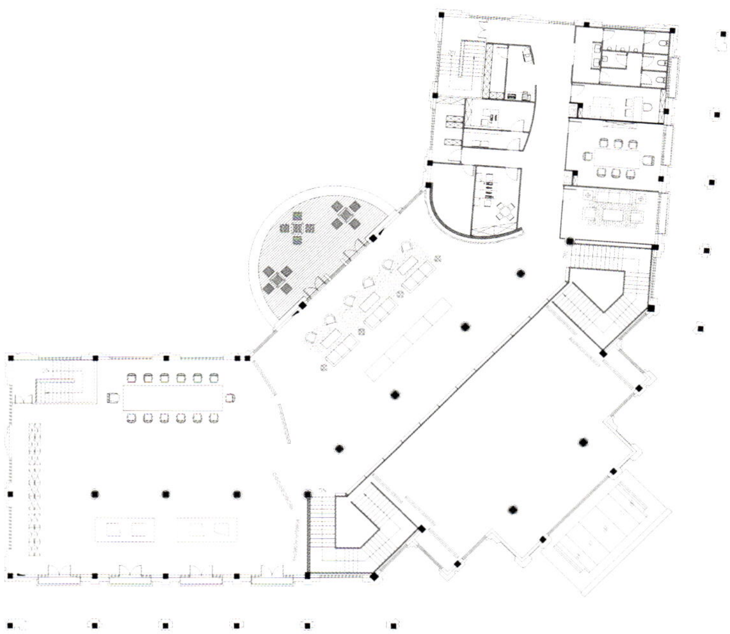

前台地面和服务台都使用白云石，极具视觉冲击力。

Dolomite are used on the reception floor and service desk, inmensely bring the visuall impact.

墙面的块面构成和光线的变化，犹如浮雕的效果。

The module cuted of the wall and the changing lignt have the effect of relief.

直线和曲线相结合，使空间富于张力。

The combination of straight and curve line make the space extending.

结合反光物料，视觉上轻盈、通透。

The combination with the reflective materials, it looks light and transparent.

白色笼罩的小局部，富于现代感。

The fraction with the covered white, endowed with the modernication.

透明材质的使用丰富了空间的视觉影像。
The transparent materials utilization enriches the visual image of space.
不规则孔状构造组成，发出淡蓝色光线。
The irregular holes stucture making up, give off the light blue light.
玻璃细柱排列的LOGO墙。
The LOGO wall lined up with the glass slender pillars.

高氏院落 Gao’s Courtyard

设计单位:高文安设计有限公司 / 设计:高文安 / 面积:1500 m² / 主要材料:石材、油漆 / 坐落地点:成都 / 完工时间:2008年11月 / 摄影:莫尚勤

Design Unit:KENNETH KO COMPANY / Design:Kenneth Ko / Area:1500 m² / Main Materials:stone、paint / Location:Chengdu / Finish Time:Nov,2008 / Photography:Steve Mok

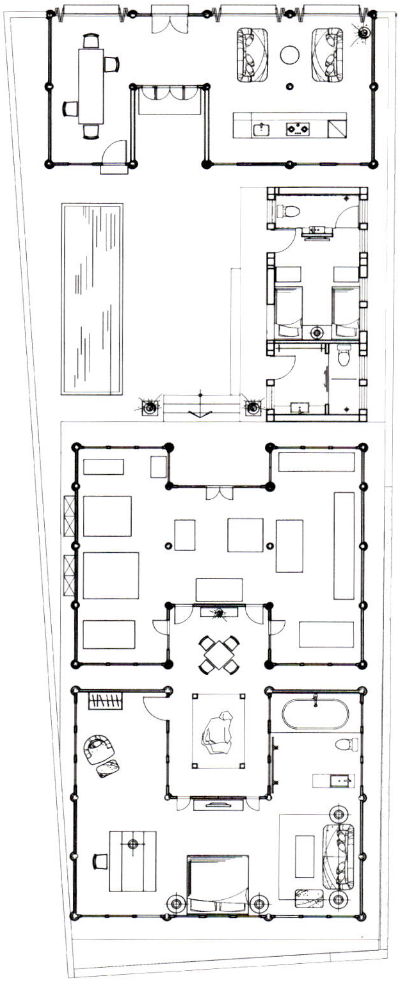

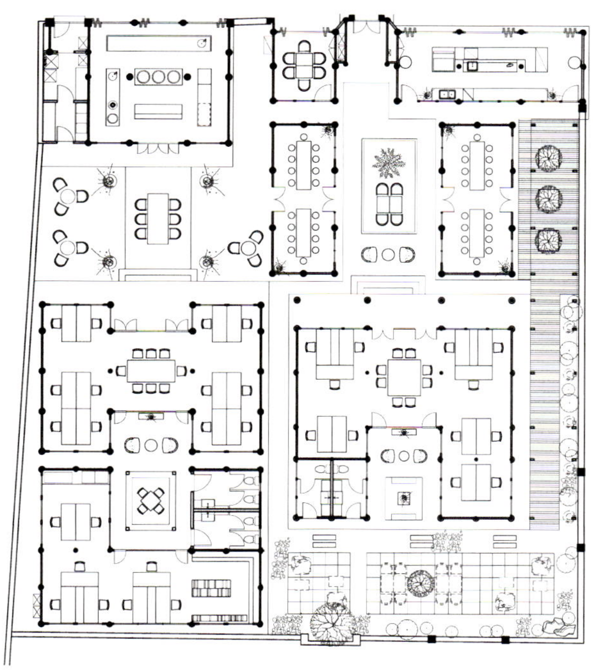

在高氏院落设计上，设计师着重于保护原有古建筑的架构和风韵，不加任何强制性的硬装，整体风格柔和低调，与宽窄巷子光影流错的韵致如出一辙。街面墙体以落地玻璃为主，意在展示自然、和谐、优雅而耐人寻味的闲适时光。

In the design of Gao’ s Courtyard, the designer has emphasized on the protection of the structure and charm of the existing building without any forcible hard decoration. The entire style looks gentle and low-profiled, just exactly same as the charm of the interweaving light and shadows in the wide and narrow lanes. The walls on the street are mainly floor glass and intend to display a natural, harmonious, elegant and thoughtful leisure time.

咖啡厅入口处的门头。

The overdoor at the entrance of the cafe.

木纹肌理的长桌古朴、自然。

The long table textured with tree grains looks simple, antique and natural.

造型奇特的椅子安放在门口，既实用又美观。

The chairs in strange models are placed in front of the door, both practical and beautiful.

沧桑斑驳的椅子，个性的画作，另类的品位。

The vicissitudinous motted chairs, unique paintings, unusual taste.

不加任何强制性的硬装，整体风格柔和低调。

It is not forced with any fixed decoration and the entire style is gentle and low-profiled.

翠竹因灯光在夜晚散发独特的魅力。

The jade green bamboos give off the unique charms at night due to the light.

在此休息让人心情平静。

People feel peaceful while resting here.

低调的具有岁月感的配饰。
The low-profiled and vicissitudinous ornaments.
老式的房梁被毫无修饰地保存下来。
The old girders are reserved without any decoration.
健身依旧是主人的最爱。
The fitness is still the favorite of the owner.

华元和庄 Hezhuang

设计单位:香港方振华设计有限公司、杭州谷谷装饰艺术有限公司 / 设计:方振华 / 参与设计:夏银苹、杨世雄 面积:406 m^2 / 主要材料:大理石、铜、墙纸、纱帘 / 坐落地点:杭州 / 工程造价:406万元 / 完工时间:2009年1月

Design Unit:Patrick Fong Design (Hong Kong) Ltd.、Hangzhou GuGu Interior Display Co.,Ltd. / Design: Patrick Fong / Participatory Design:Shirley Xia、Kevin Yang / Area:406 m^2 / Main Materials:marble、copper、wallpaper、sheer curtain / Location:Hangzhou / Project Cost:RMB 4 060 000 / Finish Time:Jan,2009

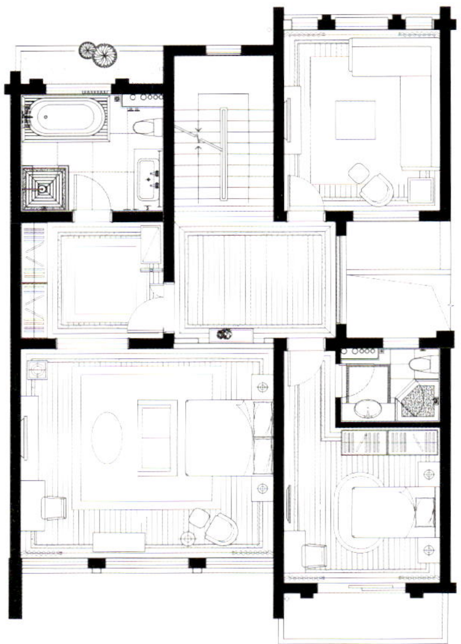

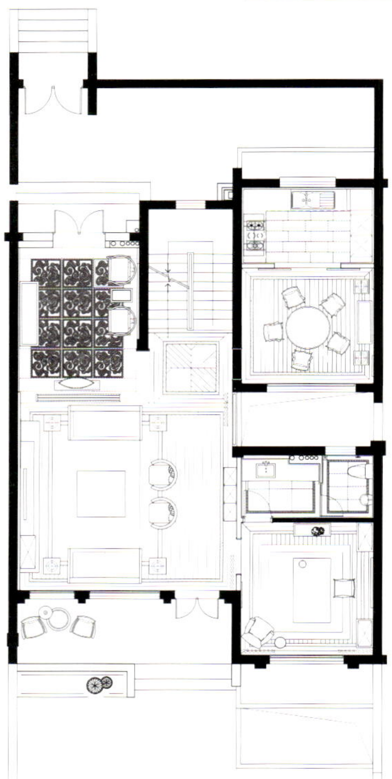

过道深处的家具充满中式味道。

The furniture in the deep aisle is full of Chinese-style atmosphere.

就餐区浑圆的灯饰象征圆满。

The perfectly round lamp decorations in the dining area symbolize the consummation.

室内设计延续建筑上的中式风格，以中国的传统文化为出发点，婉约、独特、宁静、时尚是这一户型在做室内设计时的初衷。中式建筑在现代生活中讲究“意境”，讲究“空灵”与“留白”，这点在室内色彩、材质、空间区域关系的把握和家具、装饰画、地毯等专项设计与制作上都得到了充分的体现。大气而实用的木质隔断把一层入户门厅与客厅巧妙地区分开来，美丽的艺术石雕为门厅增添了几分文化韵味，沉敛厚重的客厅以及琴棋书画样样俱全的茶室都是主人招待朋友的好地方；主卧床背景的设计透露了设计师的心思细腻，奶白色屏风在射灯的配合下显得格外引人瞩目，浅色缎面床品以及进口绒布窗帘凸显出主人房端庄华贵的气质；深色实木家具搭配暖色系床品构筑了父母房的主基调，极富心思的挂画和装饰品为它的稳重内敛增色不少；女孩房的纯白宁静、男孩房的简约时尚为房屋的小主人们实现了他们的梦想天堂。木质的稳重气质搭配新中式风格的时尚简约，沉敛深厚、文化品位浓郁，时尚感和实用性的完美结合，将中式文化的传统韵味与现代的生活方式巧妙地融为一体，隐约透露着对清雅含蓄、端庄华贵的东方式精神境界的追求。

The interior design has continued the architectural Chinese style. It proceeds from the Chinese traditional culture with the original design intention of this flat type being graceful and restrained, unique, peaceful and fashionable. The wood partition has divided the entrance hall and the living room at the first floor. The artful stone carving has added certain cultural charm to the entrance hall. The living room and the teahouse are the good places to entertain the friends; the design of the bed background in the mater' s bedroom has shown the exquisiteness of the designer. The creamy white screen is extremely attractive in combination with the spot lamps. The satin bed suite and the flannelette window curtain have prominently demonstrated a luxurious atmosphere; the solid wood furniture in combination with the warm-colored bed suite has constituted the main keynote of the parents' room. The hanging pictures and furnishings have quite increased its steadiness; the pure white and peacefulness in the girl' s room as well as the simplicity and fashion in the boy' s room have realized the small masters' dreams. The steadiness of the wood in combination with the fashion of new Chinese style have smartly integrated the traditional charms of Chinese culture together with the modern living style, which has faintly revealed the pursuit to the elegant and connotative as well as grand and luxurious eastern-style spiritual realm.

大气而实用的木质隔断把一层入户门厅与客厅巧妙地区分开来。
The grand but practical wood partitions have skillfully separated the entrance hall from the living room at the first floor.
门外的春色也不失一幅绝好的画作。
The outdoor spring scenery is definitely an extremely good picture.
至亲好友在此畅谈好不惬意。
It is a great pleasure to have free talks with the intimate friends and relatives.

紫金汇 Zijinhui

设计单位:南京远境空间环境设计有限公司 / 设计:张楠 / 参与设计:朱世艳、丰璟、张磊、张祺 / 面积: 10000 m² / 主要材料:花岗岩、大理石、紫檀木、玻璃 / 坐落地点:南京 / 工程造价:4000万元 / 完工时间:2009年1月 / 摄影:文宗博

Design Unit:Nanjing Yuan Jing Airspace Environmental Design LTD / Design:Zhang nan / Participatory Design:Zhu Shiyan、Feng jing、Zhang lei、Zhang ci / Area:10000 m² / Main Materials:granite gear、marble、rosewood、glass / Location:Nanjing / Project Cost:RMB 40 000 000 / Finish Time:Jan,2009 / Photography:Wen Zongbo

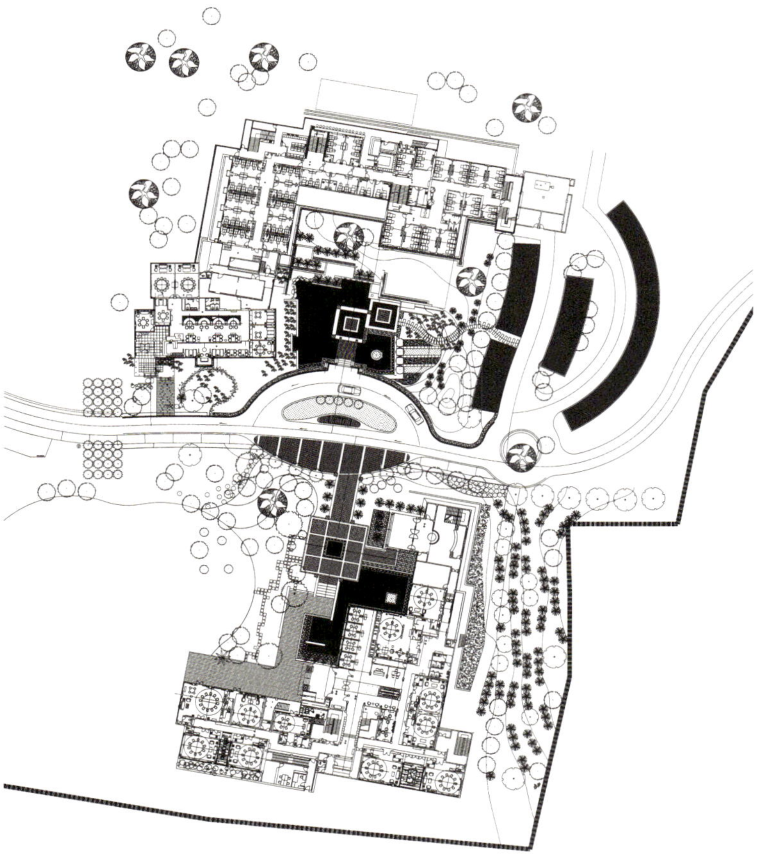

"紫金汇"由坐落在紫金山崖琵琶湖畔的建筑群改造而来，分为"明粤轩"、"琵琶湖温泉"和"印水堂"三部分。"明粤轩"针对的是高端商务和行政宴请，追求的是城墙外的精致私宅的氛围。"温泉会所"给客人提供了一个舒适私密、静心休憩的环境。而"印水堂"则在绿林掩映的湖边创造一个置身四季花园的好心情。为符合功能要求，对原有建筑进行了调整和改造。室内设计中，着重考虑了内空间和外环境的关系，对环境做了形式和意象上的"导入"；着重对大面积低矮空间进行处理，调整客人对空间的体验；合理使用材料和造型，控制成本。

Zijinhui is renovated from a building group located beside Pipa Lake at Zijin Mountain, including three parts – "Mingyue Xuan", "Hot Spring of Pipa Lake" and "Yinshui Tang". "Mingyue Xuan" is targeted at the high-level business and administrative banquets, which pursuits to the atmosphere of the delicate private residence outside the city wall. "Hot Spring Club" provides a comfortable, intimate and relaxing environment to the customers. "Yinshui Tang" creates a good mood of staying in a four-season garden beside the lake set off by a green forest. To match the functional requirements, the original building has been adjusted and modified. In the interior design, the relationship between the interior space and exterior environment has been considered with emphasis. The environment has been made with the physical and imaginary "introduction"; it has been focused on the arrangement of the large area of low spaces in order to adjust the space experience of the customers; the materials and models are used properly to control the cost.

坡形的顶与条纹状的墙面丰富了走道的景深。

The scene depth of the corridor is enriched by the wonder Sloping top and pattern wall.

色彩热烈的配饰提亮了空间。

The space is lightened by highly colorful decoration.

半弧形紫色软包既是座椅也是小隔断。

The hemicycle purple compartment forms both the chair and the small partition wall.

开放式就餐区一景。
View of the open dinning zone.
色彩缤纷的一景。
Gay colors.
仿佛艺术长廊似的走道。
Art gallery-like corridors.

天花在各种类型灯光的辉映下如行云流水。

Ceilings like floating clouds and flowing water under various lamp lights.

包间中运用了很多中式元素。

A lot of Chinese elements are applied in the compartment.

琵琶湖温泉会所入口。

The Spa Club entrance of Lake Biwa.

温泉浴区。

Spa area.

灯光赋予室内瑰丽的女性气息。
The light gives the room an woman’s elegant atmosphere.
点点蓝光从糙面石材的缝隙中透出来。
Little blue light sparks out from the gaps of the stones’ rough surface.
走道利用玻璃的透视关系扩充了空间。
The perspective effect between the glasses expands the visual field of the corridor.
灯光潜入墙壁内，形成大小不一的矩形光带。
Sneaking inside the walls ,the light forms a rectangular band of different sizes.
金色和红色主打楼梯形象设计。
The gold and red become the leading role in the staircase image design.

梦天湖商务会馆 Mengtianhu Business Hotel

设计单位:哈尔滨唯美源装饰设计有限公司 / 设计:王仲文 / 面积:5600 m² / 主要材料:大理石、柚木饰面板、马赛克、壁纸 / 坐落地点:哈尔滨 / 工程造价:1500万元 / 完工时间:2009年1月 / 摄影:张奇永、刘恒甫

Design Unit:WeiMeiYuan Design / Design:Wang Zhongwen / Area:5600 m² / Main Materials:marble、teak、mosaic、wallpaper / Location:Ha Erbin / Project Cost:RMB 15 000 000 / Finish Time:Jan,2009 / Photography:Zhang Qiyong、Liu Hengfu

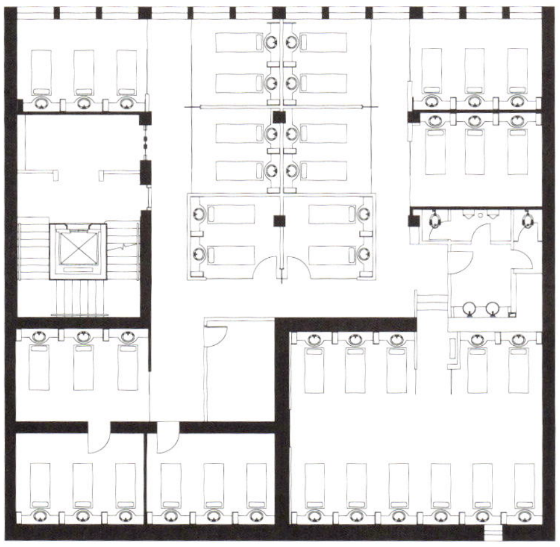

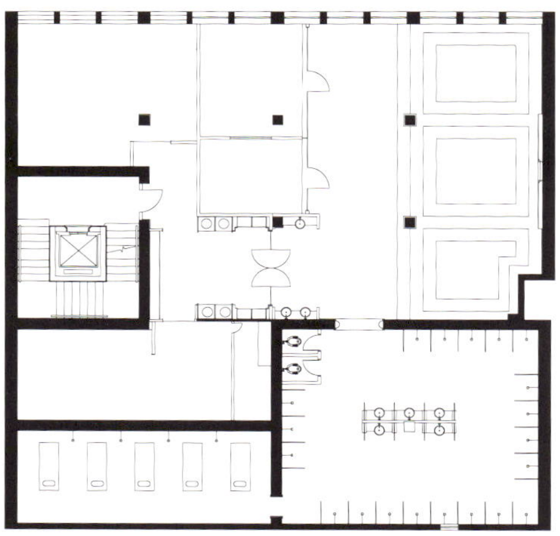

过道的处处小景雅致、温馨。
The small sights every where in the aisle look elegant and warm.
水珠的造型被运用到天花上。
The model of water drop is applied to the ceiling.

水滴般的圆环为这次设计的主题元素。

这是个以洗浴、松骨、健身合一的会馆。进入大堂，地面铺满圆形白色的玻化地砖，褐色的金砂马赛克填充其间。由于建筑原有空间很小，我们把两个侧厅同大堂相邻的墙面处理成五个理石门口，使其原有的三个厅合为一个大堂。正中对门口处设立总服务台，服务台发光的云石上贴有金银相间、大小不一的圆环同地面取得呼应，吧台一侧和下部贴满镜片，反射周围环境，增大面积。棚面退进的梯级造型之间嵌有不锈钢凹槽，增加棚面的层次。透过米黄石材的门口可以看到两侧旁厅墙面造型中的清碧山水画，增添宁静氛围。中西文化的交融，传统与现代材料的结合，塑造出一个华贵优雅、时尚稳重的空间氛围。

二层为浴区，男浴水池里蓝白相间的马赛克，延伸到墙面直到棚面。淋浴隔断的玻璃上贴满了晶莹的“水滴”同水的主题结合起来，墙面与地面饰以米白色的墙池砖，同蓝色相间的马赛克形成高调宁静的氛围。

三层的松骨区，采用褐色幻彩窗帘做成软隔断，根据功能要求的不同可灵活调整。走廊的玻璃上贴有莹光绿色亚不力的圆环，突显时尚。

It is a club integrated with bathing, bone relaxing and fitness. Upon the entry into the lobby, the ground is fully paved by the vitrified tiles with mosaic filled among them. As the space is quite small, the wall surface is arranged into five marble entrance entrances enclosed into a lobby. In the middle is a general reception desk while one side and the bottom of the bar counter are pasted with the mirrors to reflect the surrounding environment. Among the step-shape models are embedded with stainless steel grooves to increase the level sense. A picture of clean water and green mountains can been seen through the door, which has added its peaceful atmosphere. The interaction of the Chinese and western cultures as well as the combination of the traditional and modern materials have created a sumptuous and elegant atmosphere.

The second floor is the bathing area. The mosaic in the men' s bathing pool has extended to the wall till the ceiling. The glass is fully pasted with "drips" which echoes with its theme. The wall and ground are covered with the wall tiles to form a peaceful atmosphere together with mosaic.

The bone relaxing area in the third floor has applied the magic colorful curtains as the flexible partitions, which can be adjusted as per the different functions. The glass in the corridor is pasted with acrylic circles to prominently show its fashion.

规整的格局，古朴的氛围。

The regular layout, antique and simple atmosphere.

明暗有序的光影设计。

The light and shadow design in an orderly brightness and darkness.

暖黄的灯光进一步渲染了神秘的氛围。

The warm yellow light has further exaggerated the mysterious atmosphere.

透明的玻璃让空间富有情趣。

The transparent glass has made the space interesting.

浙江大酒店行政会所 Zhe Jiang Hotel Administration Club

设计单位:江苏省海岳酒店设计顾问有限公司 / 设计:姜湘岳、吴海燕 / 参与设计:纪红 / 面积:1000 m^2 / 主要材料:木饰面、石材、不锈钢、绣花皮、钢化清玻 / 坐落地点:杭州 / 工程造价:1200万元 / 完工时间:2009年4月 / 摄影:陈乙

Design Unit:Jiangsu Haiyue Hotel Design Consulting Co., Ltd. / Design:Jiang Xiangyue、Wu Haiyan / Participatory Design:Ji hong / Area:1000 m2 / Main Materials:wood veneer、stone、stainless steel、tempered glass / Location:Hangzhou / Project Cost:RMB 12 000 000 / Finish Time:Apr,2009 / Photography:Chen yi

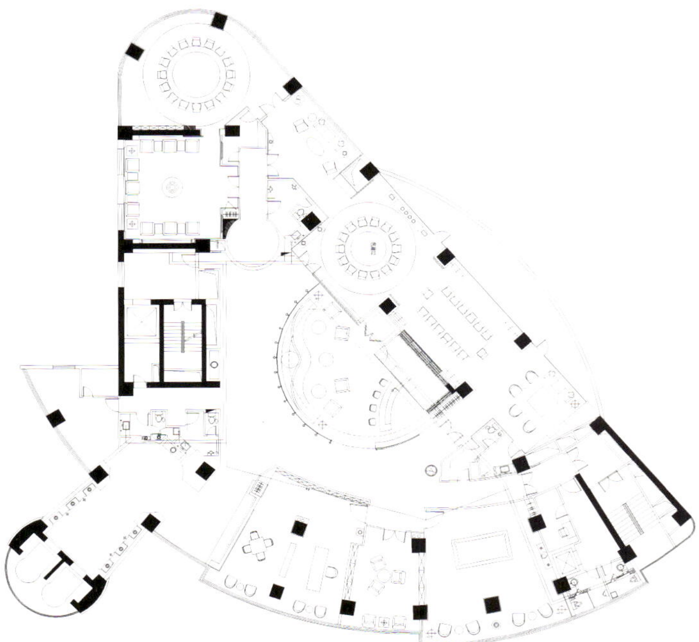

包间采用暖色调，材质、家具、壁画、色彩的选用营造了一种温馨、淡雅的气氛。
The VIP rooms have applied the warm colors. The selection of materials, furniture, wall paintings and colors has created a warm and quietly elegant atmosphere.

本案立足于营造印象自然、时尚魅力的空间设计理念，意图通过空间造型、材质、图案、艺术品的结合，营造出一个独特、时尚的空间形象。
大厅的设计给人以通透之感，顶面和地面都采用了祥云的图案，显得高贵素雅、内涵厚重。在中国，云是吉祥和高升的象征，祥云是具代表性中国文化的符号，饱含着中国传统文化的稳重大方，又不失东方感的浪漫多情；云气神奇美妙，发人遐想，令人寄思无限，可缓解客人工作一天的疲惫，身心都能得到彻底的放松。
包间采用了暖色调，其材质、家具、壁画、色彩的选用都营造了一种温馨、淡雅的气氛，是客人与好友交淡、喝茶的最佳场所。

This case is aimed at building a space idea of design that is natural in impression and fashionable and charming. It intends to create a unique, fashionable image of space through a combination of the space moulding, material, pattern and artwork
The design of the lobby gives people a transparent feeling with the top and floor bearing lucky cloud patterns, being noble, simple but elegant, rich in content. In China Clouds are a sign which is lucky and rising. Lucky clouds stand for a symbol of Chinese Culture. It fully contains the steady and good manner of the Chinese traditional culture, and may after all be accepted as romantic and full of tenderness viewed from oriental culture. Clouds look magical and beautiful and make one daydream with train of thought limitless. This can reduce a guest' s tiredness after a day' s work to relax to the full in body and mind.
VIP room in warm tones uses all material, furniture, wall paintings and colors to create a comfortable, simple but elegant atmosphere. It is the best place where guests have a talk with friends and drink tea.

墙面的做糙处理自然过渡了地面和墙面的丰富纹理。

The roughened wall surface has naturally transited the rich textures from the ground to wall surface.

低调、精细的纹案成就会议室无与伦比的内敛。

The low-profiled and delicate patterns have achieved the incomparable restraint of the conference room.

顶面和地面都采用祥云的图案，显得高贵素雅、内涵厚重。

The ceiling and ground have both applied the patterns of propitious clouds, which look noble, simple but elegant with profound meanings.

炎黄艺术馆子园 Yanhuang Art Museum — Ziyuan

设计单位:文格空间设计顾问（深圳）公司 / 设计:林文格 / 面积:800 m² / 主要材料:皮革、黑洞石、绿可木、酸枝木、巴萨黑石材、墙纸、艺术玻璃、镜钢 / 坐落地点:北京 / 工程造价:200万元 / 完工时间:2009年3月 / 摄影:林文格、吴永长

Design Unit:Wen Ge Space Design Adviser (Shenzhen) Company / Design:Lin Wenge / Area:800 m² / Main Materials:leathe、black basalt、wood、rosewood、stone、wallpaper、art glass、steel mirror / Location:Beijing / Project Cost:RMB 2 000 000 / Finish Time:May,2009 / Photography:Lin Wenge、Wu Yongchang

作为艺术馆的有机组成部分，子园是形态上的共生、风格上的延续和功能上的补充。设计师从各方面营造含蓄内敛的中式空间意境。LOGO采用“子”的篆书写法，以上下对称的镜像呈现。它演化为各处细节：地毯纹样、椅背花纹、备餐台底纹……共同烘托主题。沿用中国传统建筑模式，入口即见照壁，形似多宝格，其中摆放书简。吧台上空，黄色水晶连缀出腾龙跃动之姿。系列古典元素的组合搭配减抑商业性，突显书卷气，让内外部环境气韵浑融。

设计师以“泛空间”概念，处理内部格局。餐位设置既有中式圆桌，也有西式卡座；既有大厅散座，也有贵宾包房。以文化混搭的方式提供多重消费选择。内部雅间采用传统折叠门隔断，与外部大厅可分可合，随意调节空间大小，同时满足餐饮与聚会需求。另设吧台，夜间也可变换为酒吧。

设计思路从室内延伸到室外，将会所外的生硬围墙改建为一处跌落水景，与艺术馆正门处的景观形成对应，延续了整体建筑风格。透过落地玻璃窗，形成室内外景观互动，顾客的视线、心理得以舒展。此作品，通过文化元素的细腻运用，整体与局部空间、景观与人的行为之间气韵流动，浑然一体，自然而然与心灵产生共鸣。

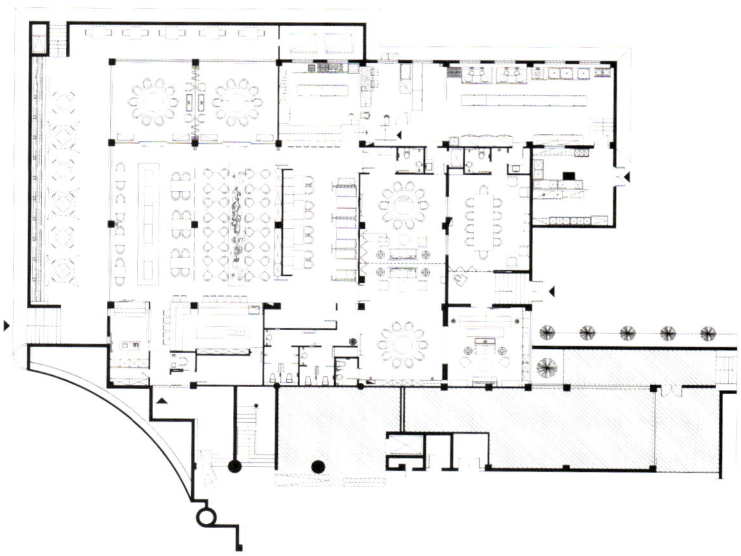

As an organic constitution of the art museum, Ziyuan is the intergrowth in form, continuity of style and supplement to the functions. The designer has created a connotative and constrained Chinese artistic conception from every aspect. LOGO has been applied with the Chinese word “子” in seal character and evolved into the details everywhere: the patterns on the carpet, flower patterns on the back of the chairs, the background pattern on the food-preparation table……the screen wall can be seen at the entrance. Above the bar counter, there is a pose of the rising flying dragon connected by yellow crystals. A match of a series of classical elements has prominently shown booklish lingering charm, thus internal breath is harmonically integrated with external one.

The designer has applied the conception of “pan-space” to handle the internal arrangement. The mulriple consumption choices are provided through the mixed arrangement. The private rooms have been applied with the traditional folding partitions to adjust the space size freely. The bar counter is otherwise arranged as a mini-bar at night.

The design philosophy has extended from the interior to the exterior to rebuild the stiff bounding wall to a water-falling sight, which echoes with the sight at the front door and forms an interaction between the interior and exterior sights through French glass windows. Through the delicate application of cultural elements, the streaming atmosphere between the entire and partial spaces as well as the sight and human behaviors, this work looks perfectly integrative and is naturally resonated with the spirits.

黑色中式餐椅搭配白色桌布、窗帘，素雅、端庄。

The black Chinese-style dining chairs go with white table cloth and curtains, simple but elegant and grand.

以文化混搭的方式提供多重消费选择。

It has applied the mixture of the cultures to provide multiple consumption choices.

竹简穿插构成的屏风别有一番韵味。

The bamboo slip alternated screen has another charm.

以“泛空间”概念，处理内部格局。
The “pan-space” conception is applied to arrange the internal layout.
另设吧台，夜间也可变换为酒吧。
The bar counter is otherwise set up and can be changed to a bar in the evening.
别致的细部配饰。
The unique decorations in details.

静水山庄 Peaceful Water Country Villa

设计单位:无锡市上瑞元筑设计制作有限公司 / 设计:冯嘉云 / 参与设计:曹小鸥 / 面积:1100 m^2 / 主要材料:仿古砖、饰面板、原木、大理石、墙纸 / 坐落地点:无锡 / 工程造价:140万元 / 完工时间:2008年12月 / 摄影:文宗博

Design Unit:Shang Rui Yuan Zhu Design Co., Ltd. / Design:Feng Jiayun / Participatory Design:Cao Xiaoou / Area:1100 m^2 / Main Materials:rustic tile、veneer、log、marble、wallpaper / Location:Wuxi / Project Cost:RMB 1 400 000 / Finish Time:Dec,2008 / Photography:Wen Zongbo

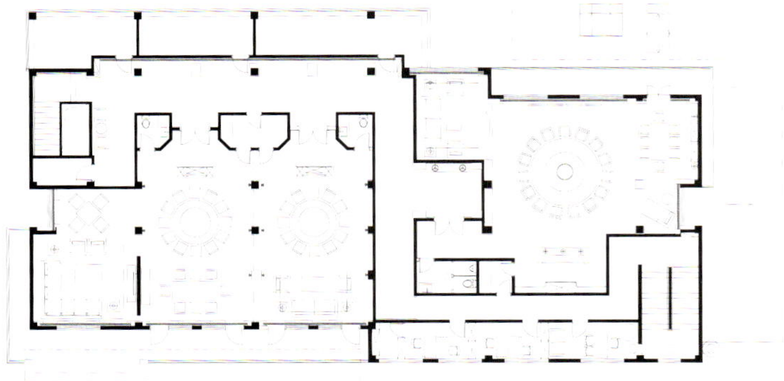

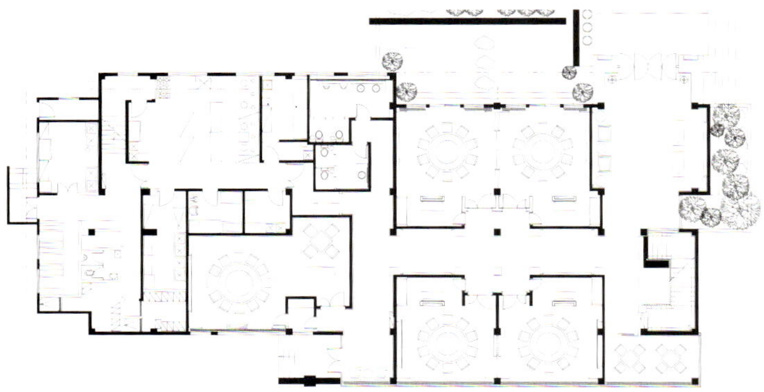

过道空间处理非常干净。

The aisle space is arranged very clean.

“拙”味十足，无需多加解释。

The “clumsy” taste is sufficient without any more explanation.

静水山庄之所以取其名，是因为它身藏花园之中，依水而建。建筑是两层的独栋小楼，有几分私家庄园的意味。于是花园别墅的情节始终萦绕，朴素、高贵、儒雅，整个设计便在心中。

内部空间的处理非常干净，顺其势，还原本来的梁柱构造，便成为设计的语言，扬长避短，自成一体，“拙”味十足，无需多加解释，作者有数，看者轻松。材料上取“朴’，把更多的用心放在表意，而非表相。细数下来并无令人称道的绝美材料、华丽造型和精湛的工艺，使用的是最朴实且平易近人的单纯素材，却通过不同质感的碰撞来产生微妙的视觉体验，又或是将未加修饰的木料作为艺术品直接呈现，还原物料的真实感。有了这些布景的衬托，家具、配饰和艺术品成为整个设计的重头戏而粉墨登场，不同的材质混搭，不同的风格混搭，不同的色彩混搭，作者不断地从不同的角度来讲述对整个设计的体会和理解。

It is a two-storey independent building with certain sense of private manor. The feeling of a garden villa has been lingering, simple 、noble、cultured. The whole design is thus in your heart.

The interior space is arranged very clean. The original beam and column structure is restored with the development of its favorable conditions and the avoidance of unfavorable ones, which is naturally integrated into a whole. “Simplicity” is taken from the materials and more attentions are emphasized on the expression of the meanings. Watching carefully, there are neither complimenting absolutely beautiful materials nor magnificent models and exquisite techniques. Only the simplest pure elements are applied, but create a delicate visual experience through the collision of different materials, or the ungroomed wooden materials are taken as the arts for direct demonstration to restore the reality of the materials. With the setoff of such sceneries, the furniture, furnishings and arts have become important points in a mixed arrangement of different materials, styles and colors. The designer is continuously describing the whole design from different points of views.

材料上取“朴”，作者把更多的用心放在表意，而非表相。
The materials are chosen as “simple”. The designer has taken more efforts on the meaning instead of looks.
落落大方的包间。
The natural and graceful VIP room.
精心挑选的室内配饰。
The elaborately selected interior ornaments.

国泉茶社 Guoquan Tea House

设计单位:南京广暄工程设计顾问有限公司 / 设计:陈广暄、马珺 / 面积:1200 m² / 主要材料:灰色木纹石、铁刀木、仿古砖、玉砂玻璃、古铜镜钢、硬底麻质绷布 / 坐落地点:南京 / 工程造价:360万元 / 完工时间:2009年3月 / 摄影:南京远景摄影设计有限公司

Design Unit:Nanjing Guangxuan Design / Design:Chen Guangxuan、Ma Jun / Area:1200 m² / Main Materials:perlino dorata Italy、cassia siamea、rustic tiles、glass、bronze mirror / Location:Nanjing / Project Cost:RMB 3 600 000 / Finish Time:Mar,2009 / Photography:Nanjing YuanJing photograph Design Co., Ltd.

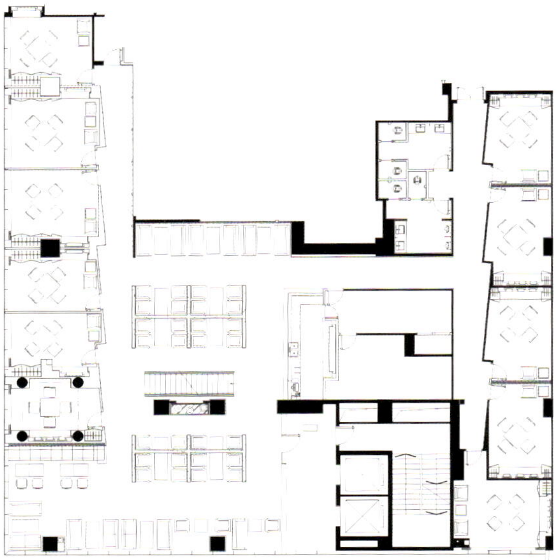

该项目位于南京玄武湖畔的国际展览中心，通过搭建形成上下两层。每层的高度都不乐观。虽然占总面积40%临窗区域及房间皆可俯瞰玄武湖，风景自不必说。但如何将余下60%无窗无景的暗区打造成准景观空间是本案的设计重点。水绿色的光效墙成为解决问题的主要手段，横向的发光方式使层高不足的问题得以缓解。另外，设计师通过若干个精巧的迷你景观穿插分布在包间区域内，打破单调平面格局。在保证私密的情况下，有的景观可以同时被走廊及包房内部观看到，产生趣味性的空间体验。最终的装饰风格是通过饰品来完成的。木质的鸟群，横卧的藤枝，高低错落的热带植物形成一种艺术化了的生态环境。

It is located in the international exhibition center beside Xuanwu Lake in Nanjing in two levels after being built. The height of each level is not so optimistic. Although in the area and rooms beside the windows with an occupation of 40% of the total area, you can overlook Xuanwu Lake and the scene is needless to say, how to develop the remaining 60% dark area without any window and scene into the quasi–sight space is the highlight of this design. The aqua light effect wall has become the mportant technique to settle the problem. The transverse illuminating method has relieved the problem of insufficient level height. Furthermore, the designer breaks through the monotonous plane layout by the alternation of numerous delicate miniature sights in VIP rooms. Some sights can also be seen both in the corridor and interior VIP rooms while keeping the intimacy to bring the interesting space experiences. The final decorative style is achieved through the ornaments. The wood bird groups, the decumbent vines as well as the tropic plants in picturesque disorder have formed an artistic ecological environment.

水绿色的光效墙成为楼梯处的点睛之笔。
The aqua light effect wall becomes the shining point at the staircase.
活泼灵动的细部。
Vivid and brilliant details.

独立而又依附整体的小包间。

The small independent VIP room, but also replying on the entirety.

水绿色光效墙依然发挥重要作用。

The aqua light effect wall still plays an important role.

四人桌喝茶、棋牌，何乐不为。

Drink and play chess or card around the four-person table, what a pleasure.

枯木横条在灯光的映照下颇有几分味道。

The withered wood and horizontal strips look quite charming under the reflection of light.

廿一会所 Twenty — One CLUB

设计单位：南京筑内空间设计顾问有限公司 / 设计:陈卫新 / 面积:2000m² / 主要材料:青砖、石材、定制木板、木门扇、玻璃、灰色镜面、织物、黑色不锈钢 / 坐落地点:南京 / 工程造价:320万元 / 完工时间:2009年5月 / 摄影:张力

Design Unit:NANJING INTERIOR DESIGN CONSULTETION CO.,LTD. / Design:Chen Weixin / Area:2000m² / Main Materials:blue brick、stone、board、wood door leaf、glass、mirror surface、fabric、stainless steel / Location:Nanjing / Project Cost:RMB 3 200 000 / Finish Time:May,2009 / Photography:Zhang li

廿一（京昆）会所位于古城南京“甘熙故宅”一侧，修复的古戏台是院落的核心。设计师在院中围绕戏台做了三面围合的浅池。晚间，冷色调的地面光与戏台之上的光影倒影相映成趣，也使戏台与观众具备了一定的观看距离，似真似幻。

考虑老宅子及院落间的通道连贯与服务功能，设计师对原建筑局部进行了改造与加建，使建筑最大程度上符合现代使用。

会所的部分建筑属于清代保留建筑，所以在这些地方采取了保护性措施，减少了对房屋本身的尺度、材质等观感的影响。

The Twenty – One Club (Jing–Kun) is situated at a side of “Former GanXi Residence” in Nanjing, an ancient stage restored being the core of the courtyard. The designer has made a theatre pit enclosed in three sides around the stage in the courtyard. In the evening the ground light with cold colour tones contracts sharply with a light and shade on the stage and it keeps a certain viewing distance between the stage and audience. It looks like a real or magic scene.

Considering the connection of passageway and function of service between ancient residence and courtyards, the designer has made a local renovation and addition to the original building, bringing it compliance with modern use where possible.

Part of the club’s building belongs to the Qing dynasty’s reserved building, where some protective measures have been taken to reduce some influence on impressions of size of the house itself and material quality.

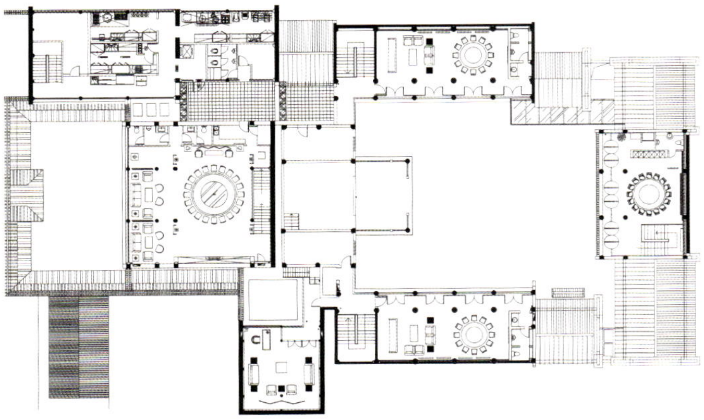

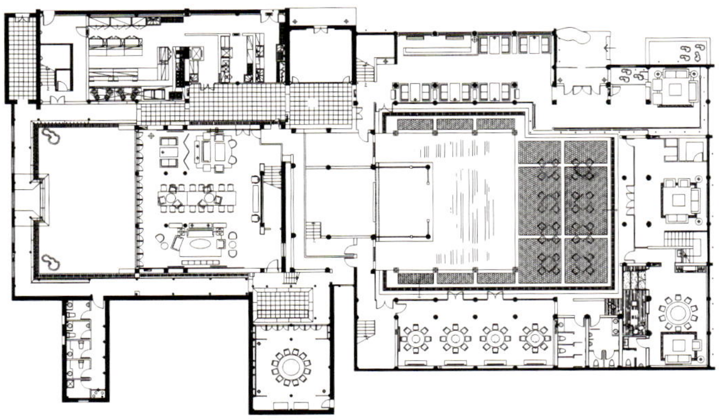

定制的精致餐盘。
Customized elaborative dinner plates.
戏台做了三面围合的浅池。
The stage is surrounded by shallow pools in three sides.
中国红色的使用与戏台色彩斑斓有内在的延续。
The usage of China red is intrinsically connected to the multiple colors of the stage.

书香世家 Scholar's Home

设计:李永强 / 面积:12000 m² / 主要材料:灰色毛面、石材、实木地板、乳胶漆、地毯、陶土砖、桃花芯木、壁纸、木皮 / 坐落地点:苏州 / 工程造价:1200万元 / 完工时间:2008年5月 / 摄影:文宗博、潘宇峰

Design:Li Yongqiang / Area:12000m² / Main Materials:stone、solid wood floor、emulsion paint、carpet、terra brick、mahogany、wallpaper、veneers / Location:Suzhou / Project Cost:RMB 12 000 000 / Finish Time:May,2008 / Photography:Wen Zongbo、Pan Yufeng

书香世家会所地理位置独特，建筑风格典雅，极具江南人文气息，建筑主体为深灰色，有较强的厚重感和时代气息。
设计沿袭以往书香世家风格，用材和表现手法上有了更大的提升。公共空间融入园林元素，叠山、流水、亭台、楼阁都有出现，力求营造苏州园林“一勺带水、一拳代山”的意境，身在其中亦能享受园林意蕴浓厚的“文人写意山水园”的乐趣。
大堂接待中心是一个挑高的空间，突出华贵与大气。内部规划合理，功能设施齐全，客人流线及服务流线清晰明朗，配合整个环境设计典雅风格，更显酒店的独特品位。

The scholar' s home club has a unique geographic location with an elegant architectural style and an extreme human and cultural atmosphere on the south bank of the Yangtse River. The main building is dark gray and has a quite strong kind and serious feeling as well as a modern atmosphere.
The design has continued the previous style of the scholar' s home, but the material application and expressive techniques are greatly upgraded. The public space has been integrated with the gardening elements such as the piled hill, running water, kiosk and pavilion etc. in pursuit to an artistic conception of "one spoon as water and one fist as hill" in Suzhou gardens. You can also enjoy the strong meaningful pleasure of "the scholar enjoys a scenic garden" that the gardens have.
The reception center in the lobby is an overhanging space to stress its luxury and grandness. The interior arrangement is reasonable with the complete functions and facilities. The guest and service streamlines are clear and outstanding and have especially shown the unique taste of the hotel in arrangement with the classic and elegant design style of the entire environment.

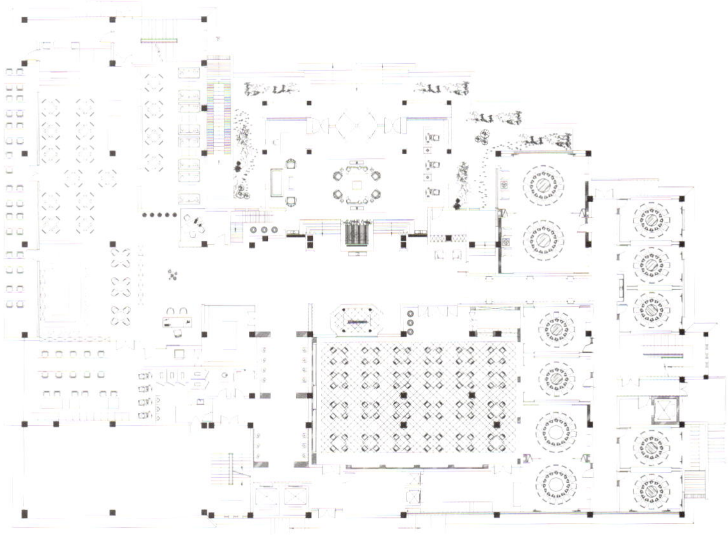

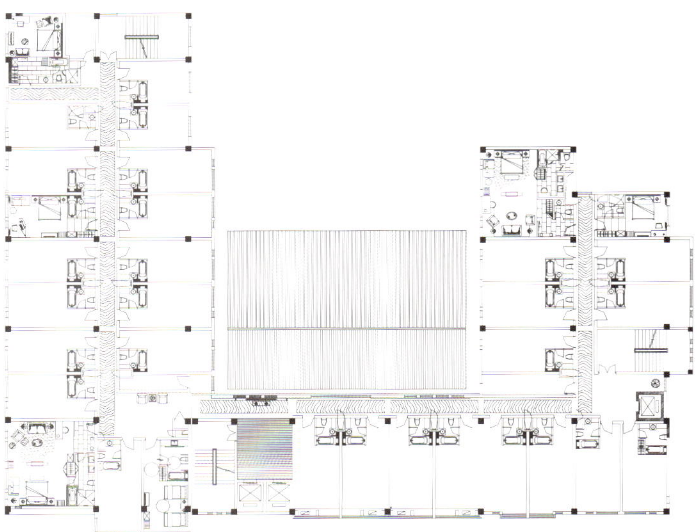

层层递进的空间关系。
The level-by-level space relationship.
大堂接待中心是一个挑高的空间，突出华贵与大气。
The reception center in the lobby is an overhanging space with stress on the nobility and grandness.

餐饮区规划合理，功能设施齐全，客人流线及服务流线清晰明朗。

The dining area is planned reasonable with complete functions and facilities. The streamlines of the guests and services are clear and distinct.

园林借景手法被借鉴。

The sight-borrowing technique in the gardens is used for reference.

意蕴浓厚的休息区。

The rest area with profound implications.

瘦西湖唐郡 Tang Manor of Slender West Lake

设计单位:金螳螂建筑装饰股份有限公司 / 设计:王方元 / 参与设计: 苏小丰 / 面积: 3000 m^2 / 主要材料:桃花芯板材、石材、织物面料、铜板、琉璃 / 坐落地点:扬州 / 工程造价:1600万元 / 完工时间:2008年12月 / 摄影:潘宇峰

Design Unit:GOLD MANTIS / Design:Wang Fangyuan / Participatory Design: Su Xiaofeng / Area:3000 m^2 / Main Materials:mahogany、stone、fabrics、copperplate、azure stone / Location:Yangzhou / Project Cost: RMB 16 000 000 / Finish Time:,Dec,2008 / Photography:Pan Yufeng

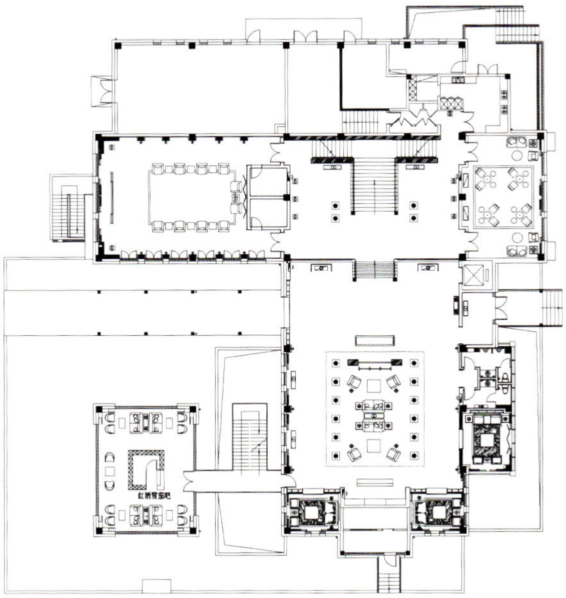

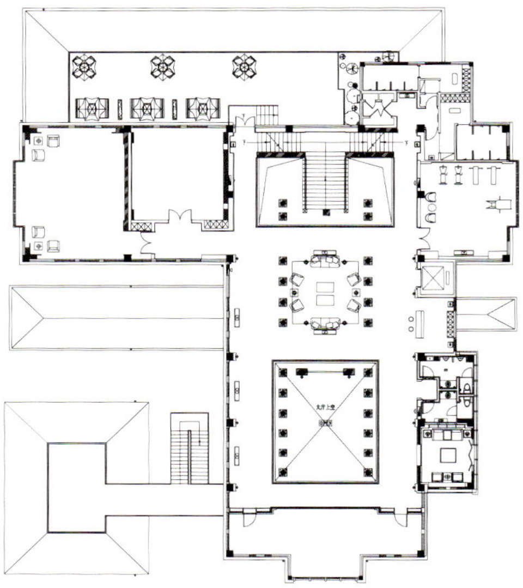

瘦西湖唐郡位于扬州瘦西湖畔。整体设计为气势宏大的中国汉唐风格。室内中性的色彩、简约的造型、巨型的体量、古朴的美感，渗透着中国古典文明的风尚。

在整体的气氛营造上，采用本木色、铜构件、纱幔、琉璃和雕花玻璃等作为主要材料；在空间形式、家具造型上，采用简洁统一的设计手法，体现自然气息和人文精神；在室内空间的组织上，各个功能区，如大堂、休息厅、红酒咖啡吧、多功能视听室、SPA区等，都是通过空间功能分布的流线关系来组织，并通过不同的家具陈设和灯光变化加以区分，充分显示空间的深度。空间之间基本上是通透、开敞的，利用雕花玻璃、中式花隔断等进行限定与围合，使室内既分割又具有一定的联系。

整体采用暖色调，自然光，人工照明采用节能暖色灯，华丽典雅、气宇非凡。利用中轴对称的方式，注意发挥结构本身的形式美，并非机械地翻版传统的符号和形式，而是运用现代的设计手法，重新演绎和定位这些传统内容，让它们有机结合。无论材质、家具、配饰、色彩都产生温馨、淡雅的气氛，让人在中式的汉唐风格中找到轻松和舒适。

Tang Manor of Slender West Lake is situated by the scenic Slender West Lake of Yangzhou City. Viewing the Tang Manor as a whole, it shows a spectacular Chinese style of Han and Tang Dynasty. Its neutral color, simple shape and huge body together appear beauty of simple and unsophisticated which prevailed in Chinese classic civilization.

The whole design of Tang Manor uses the natural wooden color, copper component, muslin and colored glaze and carved glass as its main material to create a classic atmosphere; in terms of its space form and furniture styling, it adopts a simple and uniform technique to incarnate the natural and humanistic spirit; referred to the indoor space layout, each functional area such as the lobby, lounge, red wine and coffee bar, multi-function audio-visual room and spa area and etc, are organized by the streamline relationship of the spatial functional distribution rather than only distinguished by the different furniture setting-out and the variation of the light to fully indicate the depth of the space. The carved glass and varied partitioning wall of Chinese style have defined and enclosed the space. Therefore, the different spaces are transparent and open to each other and the interior spaces are either independent or relevant with each other.

The light system adopts warm-toned natural light and the artificial light uses energy saving lamps which take on luxury, elegance and magnificence. The design uses means of axial symmetry to develop the structure' s original beauty in forms. It pays attention to apply modern design technique to re-express and define the traditional content which is integrated dynamically but not just copy those traditional symbols and forms. Whether the material, furniture, accessories or the color together create a comfy and elegant atmosphere which makes us discovering a feeling of relax and comfortable in the style of Han and Tang Dynasty.

建筑外观夜景。

The night view of architectural appearance.

入口大堂。

Entrance hall.

纱帘自上而下，摇曳生姿。

Gauze top to bottom, swaying posture of Health.

吧台将中式元素运用现代的手法展现出来。

Counter show tne Chinese elements of modern techniques.

木格外框的镜面天花丰富了空间层次。

The mirror ceiling made from wooden frame exceptionally rich the space level.

古为今用的现代中式格调。

Make the past serve the modern Chinese style.

大面积玻璃的运用让室内更接近自然。

The use of large area of glass made indoor more natural.

家具造型上，采用简洁而统一的设计手法。
Furniture design，pithiness and unity.
游泳区。
Swimming pool.
会所客房。
Club room.

首座会所 Shouzuo Club

设计单位:南京广暄工程设计顾问有限公司 / 设计:陈广暄、马珺 / 面积:1500 m² / 主要材料:有影麦哥利木、梦幻天堂大理石、手绘壁布、茶色工艺玻璃、古铜镜钢、青玉石 / 坐落地点:南京 / 工程造价:450万元 / 完工时间:2009年3月 / 摄影:南京远景摄影设计有限公司

Design Unit:Nanjing Guangxuan Design / Design:Chen Guangxuan、Ma Jun / Area:1500 m² / Main Materials:makore、marble、wallhanging、technology glass、bronze mirror、green onyx-B / Location: Nanjing / Project Cost:RMB 4 500 000 / Finish Time:Mar,2009 / Photography:Nanjing YuanJing photograph Design Co., Ltd.

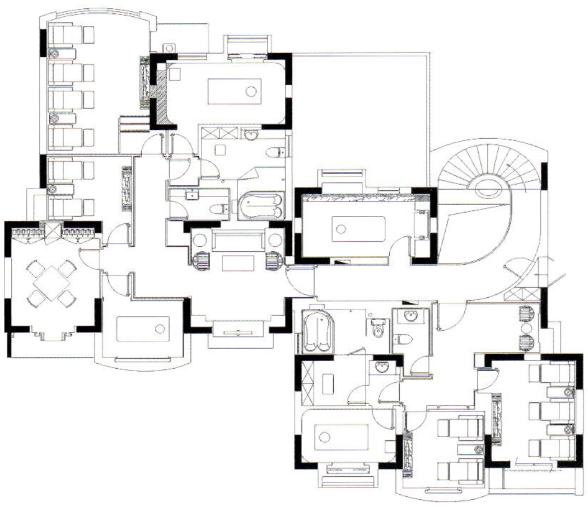

这是一个由室内设计师、平面设计师、文联名家、书画名家共同参与完成的设计项目。对于“中式风格”这样一个大的主题，每位参与者都有自己不同的视角与理解。对于一个只有1500m²的项目来说，空间中所能承载的元素是有限的。最终大家选择了“印章”作为文化含义的切入点。由项目的案名“首座”开始就以一枚闲章形式展现给来客。所有包间及主要区域的天花都是以纯白的平面吊顶配以一方闲章图案来完成的。印章给了留白空间以注解，留白给了印章涵义以想象。所有包间都选用词牌名冠名，包间天花一角的闲章内容则是出自该词牌里的佳句，而以该佳句的汉字字体本身来创作的包间背景图案则是一次独特的尝试。如“一蓑烟雨任平生”的汉字经过组合变作了一个蓑翁独钓寒江诗意画面。平面设计师因为摆脱传统书画创作条框，而极大地提升了创作热情。或许中国的文化精神早已凌驾艺术手段之上，本就不必拘泥于表现形式。在这里是“字”抑或“画”还是“留白”触动人们的审美情结也无需去争论了吧。

It is a design project jointly participated and done by interior designers, plane designers, the famous experts from China Federation of Literary and Art Circles in painting and calligraphy. Finally, they select “stamp” as the starting point of the cultural meanings. “Shouzuo” is firstly displayed in a form of spare stamp. The ceilings in all VIP rooms and main areas are accomplished by pure white plane ceiling together with a spare stamp figure. The stamp has interpreted the blank space while the blankness has endowed the stamp meaning with imaginations. VIP rooms are named after the titles of ancient poems. The contents in the spare stamp a one corner of the ceiling in VIP room comes from the well-worded phase in this ancient poem. It is a unique attempt to create the background figure of a VIP room by the well-worded phase and the Chinese characters. For example, the Chinese characters of “it is enough for me to stay in the bad weather with a coir raincoat” have become an artistic and poetic picture that an old man with a coir raincoat is fishing alone beside the cold river after the arrangement.

入口处梦幻天堂大理石与LOGO形成鲜明对比。
The dreaming heavenly marble at the entrance has set off sharply against the LOGO.
旋转扶梯曲线灵动。
The curves of the handrail of eddied ladder look brilliant.

太师椅静坐，古韵十足。
The fauteuil is standing there quietly with full antique charm.
竹叶的绘制意味深长。
The bamboo leaves are drawn meaningfully.
背景即是一首诗、一幅画。
The background is just a poem, a picture.
镜面反射的温泉小景。
The spring water scene reflected by the mirror.

沿河人家 Riverside Home

设计单位:无锡市上瑞元筑设计制作有限公司 / 设计:冯嘉云 / 面积:300m² / 主要材料:老砖、原木、玻璃、墙纸 / 坐落地点:无锡 / 工程造价:60万元 / 完工时间:2008年8月 / 摄影:文宗博

Design Unit:Shang Rui Yuan Zhu Design Co., Ltd. / Design:Feng Jiayun / Area:300m² / Main Materials: old brick、log、glass、wallpaper / Location:Wuxi / Project Cost:RMB 600 000 / Finish Time:Aug,2008 / Photography:Wen Zongbo

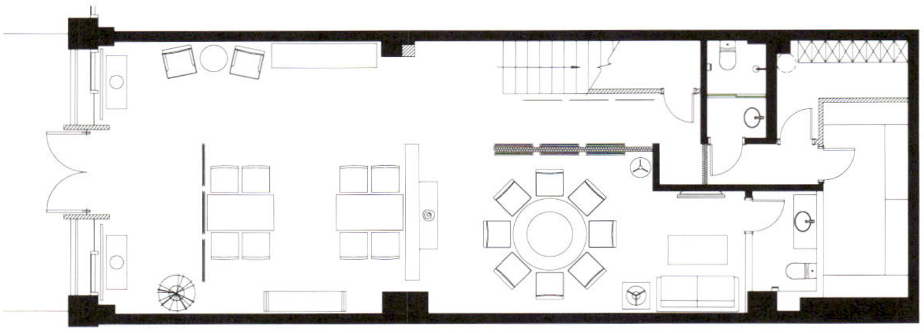

老木头在装修中让人在不经意中对时空有所回味。
The old woods in the decoration have reminded you of the old time and space.
旧家具和艺术品的配置使餐厅有了人文气息。
The arrangement of the old furniture and arts has made the restaurant with more human cultural atmosphere.

沿河人家是一家小型精致餐厅，客户的定位要求很高，由于资金有限，对设计是一个很大的制约，但同时激发了设计的思路开拓。由于本身的经营定位和经费限制，所以设计必须尽量使用廉价的材料，来达到设计所要求的气质，并尽量做到少做多放，使空间以艺术品和家具的精神气质来表现。砖和老木头在装修中让人在不经意中对时空有所回味，旧家具和艺术品的配置，也使餐厅更添了人文气息。整体设计就是化腐朽为神奇，在怀旧朴实中体现典雅的空间气质。

Riverside home is a small delicate restaurant with high requirements orientated by the customers. The limited capital is a big restriction to the design, but it also inspires the development of the design ideas. Due to its business orientation and the limited fund, the design shall use the cheap materials to the maximum to reach the disposition required by the design and try to do less and put more so that the space can be demonstrated through the spiritual dispositions of the arts and furniture. The bricks and old woods in the decoration have reminded you of the old time and space. The arrangement of the old furniture and arts has made the restaurant with more human cultural atmosphere. In a word, the design of riverside home is to change the decrepitude to magic and reflects the elegant space atmosphere through reminiscence and simplicity.

楼道上点点烛光浪漫、优雅。

The dotted candle light in the corridor looks romantic and elegant.

化腐朽为神奇，在怀旧朴实中体现典雅的空间气质。

Change the decrepitude to magic and reflects the elegant space atmosphere through reminiscence and simplicity.

空间以艺术品和家具的精神气质来表现。

The space is demonstrated through the spiritual dispositions of the arts and furniture.

勺子拼成的花朵很有心意。

The flower assembled by the spoons is very fresh.

包间沉稳大气。
VIP room looks calm and grand.
家具和艺术品在细节处见真功夫。
The furniture and artworks shows the craftsmanship in their details.

Infusion lunge Infusion lunge

设计单位:香港德坚设计公司 / 设计:陈德坚 / 面积:605 m² / 主要材料:钢、木材、玻璃、塑料喷涂、金属 / 坐落地点:旧金山 / 工程造价:500 万元 / 完工时间:2009年1月 / 摄影:陈德坚

Design Unit:Kinney chan and associates / Design:Kinney chan / Area:605 m² / Main Materials:steel、wood、glass、plastic spraying、coherer / Location:San Francisco / Project Cost:RMB 5 000 000 / Finish Time:Jan,2009 / Photography:Kinney chan

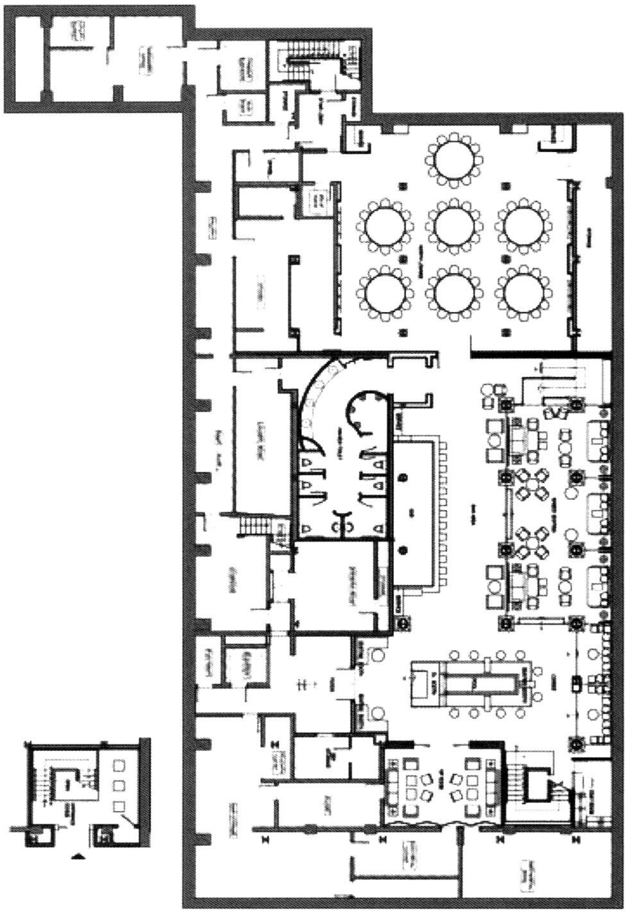

精致的格窗基于中国传统建筑文化精髓。
The delicate pane windows are based on the essences of Chinese traditional architectural cultures.

“对于酒吧设计，要了解当地酒吧文化，有音乐、有酒、还有很多的人，一般对于酒吧的认识似乎只至于此，但实际上，各地还是有所不同。不同的国家有不同的传统及文化，但唯一相同的就是给人有一个聚会的地方，也可以一个人喝酒，在黑暗的光线里，品味着一天的心情。”德坚设计总监陈德坚娓娓道来。

的确，有着15年的室内设计经历，50多间酒吧设计经验的香港设计师陈德坚，从动态到静态无不诠释酒吧的灵魂。而对于2009年1月刚竣工，美国加利福尼亚州旧金山的Infusion lunge，则是对酒吧的另一种表达。

本案坐落于旧金山市中心的联合广场，设计师根据地域文化及空间的需求，打造一个亚洲最具灵感酒吧和餐厅的多功能空间。6500 m²的空间涵盖了功能房间区、酒吧区、贵宾区。整个空间以敞开式格局呈现，“外圆内方”的哲学含义，在此设计中也发挥得淋漓精致。设计中基于中国传统建筑文化精髓，体现中西合并的空间气氛。入口处，从若干拱形门进入不同的区域。大量的红色布满整个空间，中式的器皿、饰品，与家具形成两种动律的对比；精致的格窗，镂空的廊柱，突出了空间的复古与轮廓感。贵宾区，注重奢华与高贵，硕大的LED照明系统照亮蚀刻玻璃栏。黑色壁柜上镶嵌红色庄严的“双龙戏珠”很是应景。功能房间区，中国北方的“土炕”成为空间的设计重点，方形、圆形的绸缎靠垫，搭配圆形的茶几，三五人小憩，很具韵味。墙壁上运用江南风格的瓦片，深色金属鱼鳞片状的隔门，既具有中式的复古风格，也不失现代感的舒适。

在富丽堂皇的接待处旁，由拱门可进入盥洗室，马赛克的瓷砖镶嵌的整个墙壁上，勾勒出动感的曲线，黑色的大理石台面和白色盥洗盆，和谐搭配出安静的透明效果，其间大量的金黄色又增加空间感，丰富了色彩。

Infusion lunge的奢华与复古的中式酒吧餐厅，在设计上的突破与创新，不仅是设计师自我的展示，亦是亚洲的启发而影响到欧洲乃至世界的震撼。“好的设计师，也是一个好的艺术家。”Infusion lunge正在演绎......

Located in the popular Union Square beneath the Fusion Hotel in Downtown San Francisco, Infusion Lounge is an Asian-inspired bar and restaurant executed by Hong Kong's hottest designer, Kinney Chan, will offer a full menu of Cal-Asian fusion small plates paired with exquisite ambiance and exceptional service. The lounge will also redefines an upscale nightlife not just for San Francisco, but for the industry itself.

The 6,500-square-foot space features a main room with a full bar, an elevated VIP area, and striking latticed columns based on Chinese architecture and window patterns. In the second room, which will feature a second DJ stage and will also be available for banquets, fashion shows, and other events, a programmable LED lighting system illuminates the etched-glass columns. The most arresting area of the club is the unisex restroom: The tile mosaic on the walls curve around the sinks and stalls, with high-tech Dyson Airblade hand dryers as a main attraction on the way out.

A combination of first rate service and extravagant ambiance, consisting of chic- innovative design, is sure to make an impact on Bay Area bar scene. This sophisticated, Asian inspired, sub level lounge will become a must-see destination.

镂空的廊柱，突出了空间的复古与轮廓感。
The hollowed-out corridor columns have stressed the antique and profile of the space.
中式的器皿、饰品，与家具形成两种动律的对比。
The Chinese household utensils and ornaments as well as the furniture have formed two kinds of dynamic contrasts.
由拱门可进入盥洗室。
Enter the washroom through the archway.
墙壁上运用江南风格的瓦片。
The tiles in the style on the south of the Yangtse River are applied to the wall.

静SPA Quiet SPA

设计单位:哈尔滨唯美源装饰设计有限公司 / 设计:刘志 / 面积:2000 m² / 主要材料:竹子饰面板、质感涂料、马赛克 / 坐落地点:郑州 / 工程造价:2000万元 / 完工时间:2008年10月 / 摄影:张奇永

Design Unit:WeiMeiYuan Design / Design:Liu zhi / Area:2000 m² / Main Materials:bamboo veneer、texture coating、mosaic / Location:Zhengzhou / Project Cost:RMB 20 000 000 / Finish Time:Oct,2008 / Photography:Zhang Qiyong

自然与水永远是SPA的主题，在“静 ”的设计中将自然与水的元素提炼出来，用不同的手法表现。当您走进“静”要踏水而过时，鱼儿在脚下游动，墙面的造型如波浪一样飘逸，抬头望去，天棚如映着阳光的水滴。不同形态的水和细细的水声，让人的心绪变得平静。自然的元素表现在设计符号和用材方面，我们将大自然中花瓣和叶子的形状结合设计出简洁明朗的符号，用不同的材质形式融会到空间中，装饰材料也选择像竹子面板和质感涂料等贴近自然的，让人充分体验到SPA带来的放松身心的感觉。

Water and nature are the theme of SPA for always. Various methods are used in the designing of "Quiet" to purify the elements of nature and water. As you are to wade, you can see the fish swimming beneath your feet, the model on the walls is as elegant as waves. When you look up at the ceiling, it looks like water drop. Different forms of water and its murmuring quietens your nervous. Nature elements are embodied in the symbols designing and materials adoption. Combining the forms of leaves and petals in nature, we create the brief symbols. We blend different material qualities into the space. For the decoration materials, we also chcose those are close to nature, like bamboo board and paintings, which can make you thoroughly feel relax brought by SPA.

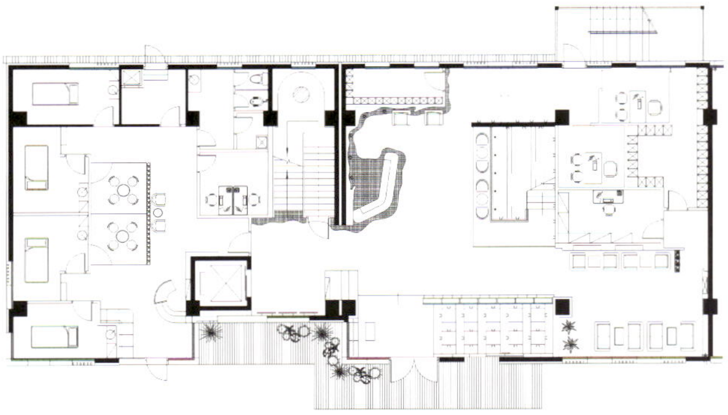

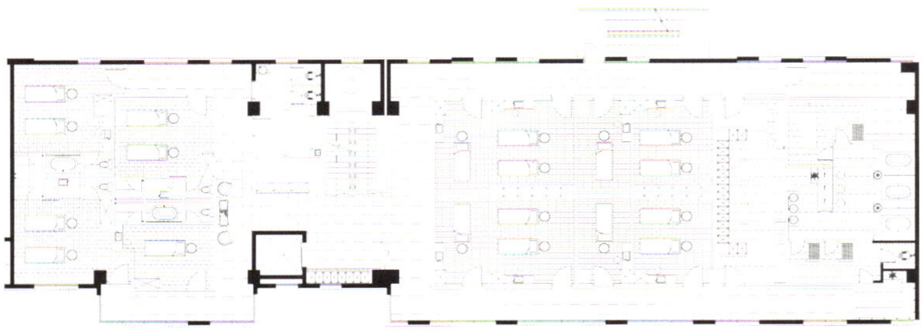

会所外观。
Visual appearance of the club.
天棚如映着阳光的水滴。
The ceiling looks like the water drips reflected by sunlight.

磨砂玻璃构成的门。
Frosted glass door.
按摩区。
The massage area.
安静的过道。
The quiet corridor.
会所一角。
One corner of the club.

SOKO美容院 SOKO Beauty Salon

设计单位:哈尔滨唯美源装饰设计有限公司 / 设计:韩冠恒 / 面积:300 m^2 / 主要材料:水泥、地坪漆 / 坐落地点:哈尔滨 / 工程造价:200万元 / 完工时间:2008年6月 / 摄影:张奇永

Design Unit:WeiMeiYuan Design / Design:Han Guanheng / Area:300 m^2 / Main Materials:cement、flooring coating / Location:Ha Erbin / Project Cost:RMB 2 000 000 / Finish Time:Jun,2008 / Photography:Zhang Qiyong

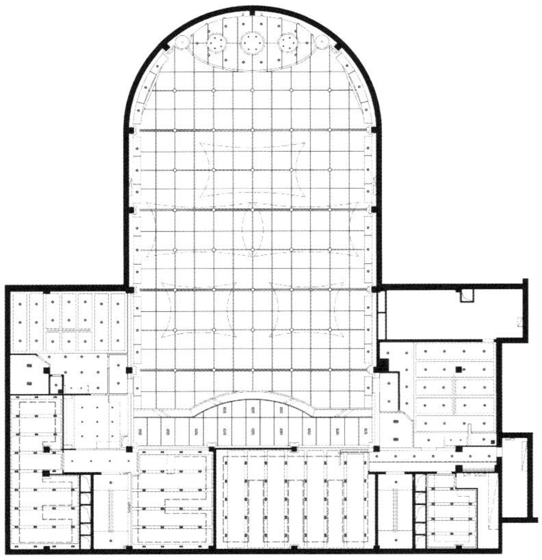

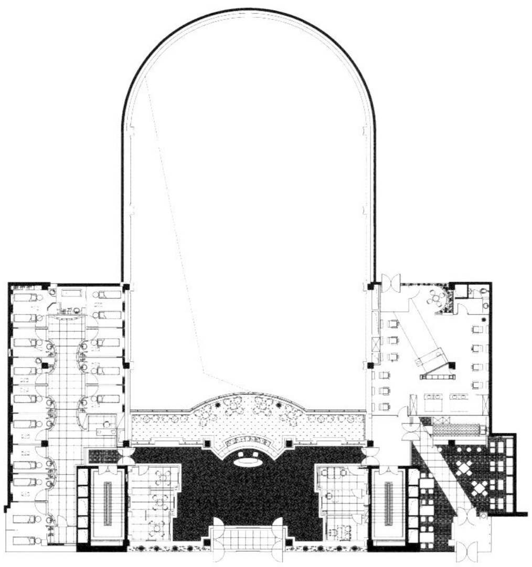

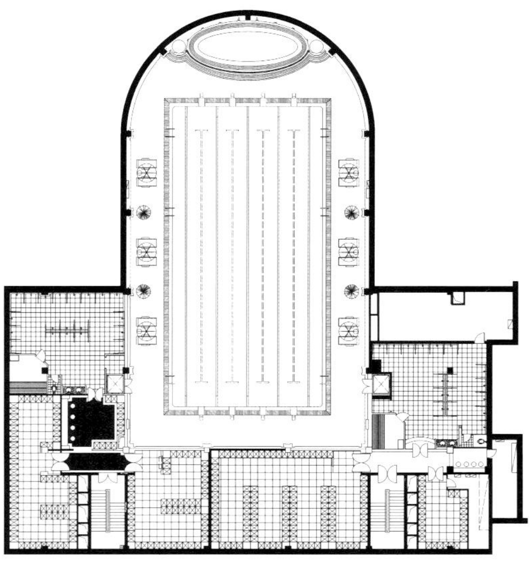

设计风格着重体现中国文化内涵的宗旨，用古典文样做设计语言，以黑、红为主基调。更以裸露的顶棚及设备管线等有规律的排列，表现出空间的运动和健康的感受。运用体块的穿插、墙体的转折、黑红两色穿插使用的方法来划分空间，体现出力量感。同时不锈钢的万字格、黑檀板造型和热带雨林墙纸烘托出古典文化中的静、雅、素，营造具有人文情怀的健身环境。

主楼是U型的四层综合楼，环抱游泳馆。前厅用黑色转折的墙体和万字格不锈钢造型穿插形成两侧的洽谈区。黑檀板扇状形象墙正对主入口，一排大尺度的沙灯将客人引向前台。形象墙后面是加建的休息区，大面积落地玻璃窗将其与泳馆分割开，使前厅空间得以延续。前台两侧是空间的主通道，把客人引入其他功能空间。美容美发区用碎瓷砖在墙面拼贴出河流的脉络，同黑檀板造型、鹅卵石条砖演绎出美妙空间。咖啡厅黑胡桃的木格天花与地面条型砖遥相呼应，对比鲜明的深浅造型交叉分割出吧台区和卡座区，更以不同形式的桌椅来细化就餐区。

二、三层为健身区，裸露的天花、红色的条板，既保证了视觉的通透性，又与整个气氛融合，黑色墙面同红色烤漆玻璃共同塑造出一个富有生命力的健身氛围。

The style has continued "Queen" to reflect the tenet of Chinese cultural meanings with black and red as main colors. Moreover, the exposed ceiling and equipment pipelines etc. are lined in order to show the sportive and healthy feelings in the space. The stainless steel grid in Chinese character "Wan" and the ebony panel as well as the wall paper of tropic rain forests have set off the "quietness", "elegance" and "simplicity" in the classic culture to create a fitness environment of human and cultural feelings.

The main building is a four-storey U-shaped comprehensive building encircling the swimming hall. The black turning wall and the model of stainless steel grid in Chinese character "Wan" are applied in the antechamber to cross each other to form the discussion areas on both sides. The black walnut wood grid ceiling and the strip-shaped tiles on the ground remotely echo with each other in the cafe. The dark and light models with bright contrast have intersected into the bar area and booth area. The tables and chairs in different forms are used to subdivide the dining area.

The second and third floors are used for the fitness area. The exposed ceiling and strip panels both ensure the visual transparency and are integrated with the whole atmosphere. The black wall together with the red lacquered glass has created a vigorous fitness atmosphere.

静、雅、素，营造具有人文情怀的健身环境。

"Quietness", "elegance" and "simplicity" create a fitness environment of human cultural passions.

仿佛太空舱一样的入口。
An entrance like a space capsule.
进门小阶梯。
Small access staircase.
一桌一椅一世界。
One table, one chair and one world.
水滴般的镜子很柔美。
The drip-like mirror looks gentle and beautiful.

天京会所 Tianjing Club

设计单位:浩澜设计事务所 / 设计:李浩澜、曹端顺 / 面积:2500 m² / 主要材料:镜面、墙纸 / 坐落地点:南京 / 工程造价:500万元 / 完工时间:2009年8月 / 摄影:王刚

Design Unit:HaoLan Design Office / Design:Li Haolan、Cao Duanshun / Area: 2500 m² / Main Materials: mirror surface、Wallpaper / Location:Nanjing / Project Cost:RMB 5 000 000 / Finish Time:Aug,2009 / Photography:Wang gang

酷而有形、深邃而妩媚，是本设计给人最强烈感受。在建筑空间格局受限的前提下，体现豪华感、档次感，是本设计着重体现的主题。设计师通过刚性与柔性的综合演绎，空间与材料的混搭，运用玻璃、不锈钢、地毯、皮革、布艺等各种材料，结合光效，赋予空间蓝色的神秘感，体现了奢华气派的氛围。诱惑的灯光，女人的眼神，这便是热烈、舒张、含蓄、妩媚并蕴含着神秘色彩的性感空间。

As to this design, what impress you most must be its cool, mysterious and charming style. To pursue luxury and top grade in a space with a rather limited size and layout is the objective of this design. The designer well demonstrated the combination of softness and hardness, mixed and matched various materials, including glass, stainless steel, rugs, leathers and fabrics etc., utilized lighting, and finally created a mysterious and luxurious atmosphere. With the tempting light and women' s glance, we can obviously feel the fierce, relaxing, implicit or charming aspect of this sexy and mysterious space.

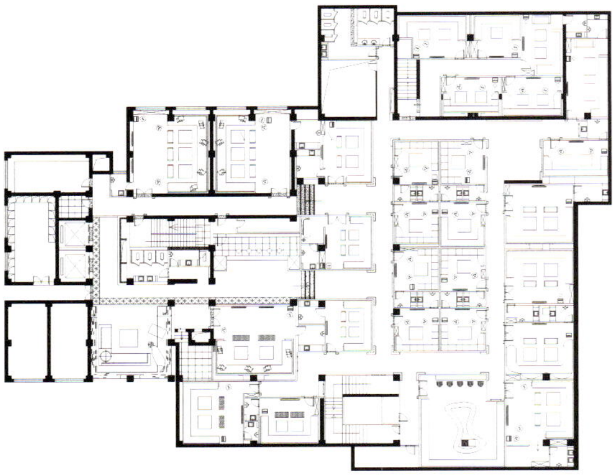

酷而有形、深邃而妩媚。
Cool and visible, profound and charming.
结合光效，赋予空间蓝色的神秘感。
The light effects are combined to endow the blue mystery of the space.

E·G造型沙龙 E·G Style Salon

设计单位:汉文设计 / 设计:万宏伟 / 面积:600 m^2 / 主要材料:爵士白、深蓝灰色钢琴漆、白色砖墙、定制刻花
坐落地点:宁波 / 工程造价:150万元 / 完工时间:2008年12月 / 摄影:杨旭

Design Unit:Han Wen Design / Design:Wan Hongwei / Area:600 m^2 / Main Materials:ajax、piano lacquer、white brick wall、engraving / Location:Ningbo / Project Cost:RMB 1 500 000 / Finish Time:Dec,2008 / Photography:Yang xu

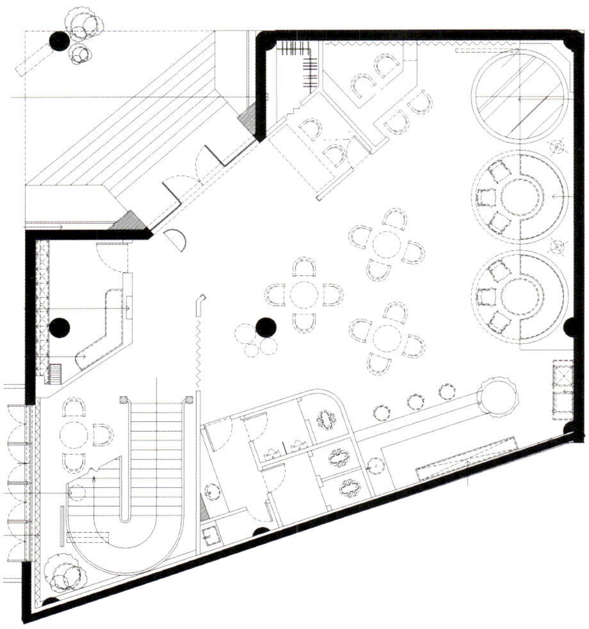

对空间性质的转换与引导成为设计重点，二个楼层的店务空间，一楼空间规划为多功能造型艺术沙龙区及美容美甲区域，二楼为专业美发区域。对空间及风格的整合，对色彩及灯光的调控有更多的思考。典型的ART DECO 艺术装饰风格，结合当代艺术摄影，体现优雅从容的做派及高端专业的发型造型空间。

The conversion and guidance to the space features have become the key design point. There are two floors of shop space: the space at the first floor is designed into a multifunctional style art salon area and a hairdressing and nail beatifying area; the second floor is a professional hairdressing area. More consideration is taken on the rearrangement of the space and style as well as the adjustment and control of the colors and light. The typical ART DECO style in combination with the contemporary art photographing has demonstrated the elegant and leisured practices as well as a superior professional hairdressing space.

秩序井然的洗头区。
The well-ordered shampoo area.
入口狭长的过道。
The narrow long corridor at the entrance.

E·G造型

KÉRASTASE

白色砖墙简洁、大方。

The white brick wall looks simple and decent.

造型别致的小灯。

The unique small lamp.

典型的ART DECO艺术装饰风格，结合当代艺术摄影，体现优雅从容的做派及高端专业的发型造型空间。

The typical ART DECO style is combined with the contemporary artistic photography to reflect a superior professional hairdressing space in an elegant and leisurely manner.

SIN酒吧 SIN

设计单位:柯凯建筑设计顾问(上海)有限公司 / 设计:Andrea Destefanis、Nora Von Nordenskjoeld、杨璇、熊来胜、陈颖 / 面积:1700 m² / 主要材料:石材、竹木地板、黑漆饰面木板 / 坐落地点:上海 / 完工时间:2008年11月

Design Unit:Kokaistudios / Design:Andrea Destefanis、Nora Von Nordenskjoeld、Yang xuan、Xiong Laisheng、Chen ying / Area:1700 m² / Main Materials:stone、bamboo flooring、veneer / Location: Shanghai / Finish Time:Nov,2008

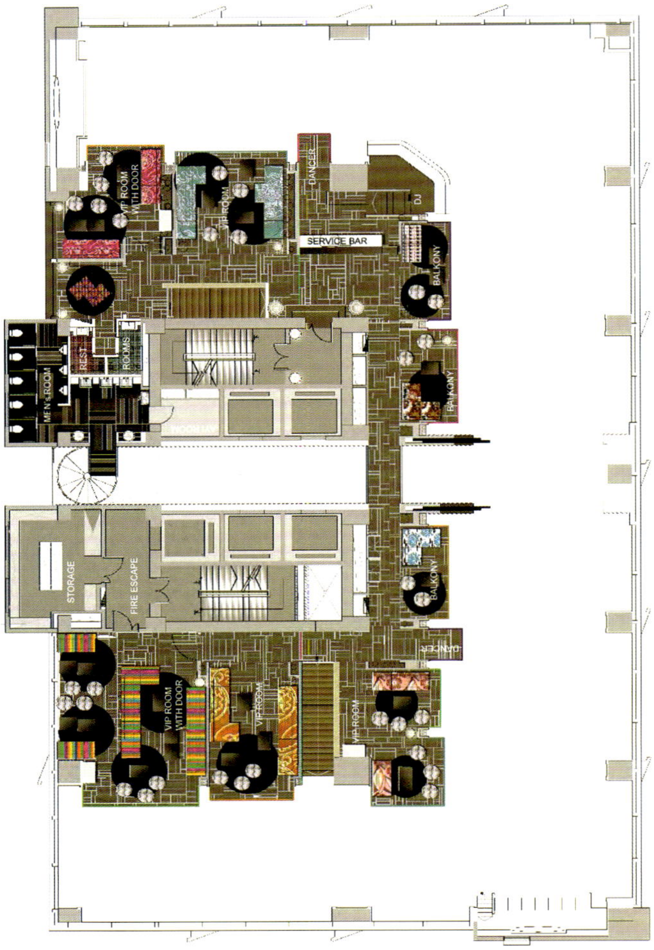

项目设计参照《圣经》里的七种原罪，定义了室内的七种主色调，并贯穿于整个项目之中，各成风格。采用自然形态和材料，营造集合海洋和岛屿、山川和洞穴、太阳与风这些对应概念风貌的布局。

为体现伊甸园自然意境的风貌，地面采用纹理精美的石材反射光线，给人海洋起伏流动之感，竹木搭造而成的休闲区点缀“海面”之上。墙面均采用漆黑色调，唯有核心墙的饰面以三维立体的木制纹理表现岩石的质感。同时引入亚当和夏娃故事中标志性的元素，做成室内装饰装置，如天使翅膀、红苹果、蛇，经过深入研究蛇的表皮结构和它的动态行为，最终在入口墙面上用了模仿蛇皮的特别定制设计——菱形不锈钢片和天鹅绒拼接而成的饰面。

在空间布局方面，有意通过隔断和特别装饰元素使空间结构变得复杂，营造一种心理上私密的酒吧氛围。内部人流动线设计意在引导人们以探索的方式穿梭于不同区域。厨房设备等服务区域被隐藏在夹层结构一层。主入口过道两边设计了超大夸张的移门，将SIN分成两部分，满足波动的客流情况。内部空间设计分隔成多个更小区域，具备更强的灵活性。

从顶层电梯出来，过道两边以金属质感的仿蛇皮纹理墙面不断折射室内光线，营造纵横交错的效果。过道铺设的红色地毯上方悬挂着通长23m，用1.7万枚玻璃水晶定制成的蛇形吊灯，将视线引向入口处的标志。

This design has been referred to seven original sins in the bible to define the seven main interior colors and run throughout the whole project in different styles. It has created an overall arrangement of the ocean and island, mountain, river and cave as well as sun and wind corresponding to the conceptual views.

To show the view of Eden, the floor has been applied with the stone materials to reflect the light. The wall is all painted black except the veneer of the core wall, which reflects the quality of the rocks through the three-dimensional wood textures. On the wall at the entrance, a specially customized design of imitating the snake skin – the veneer assembled by rhombic stainless steel pieces and velvets – has been applied.

In the space arrangement, it intends to complicate the space layout through the partitions to create an intimate bar atmosphere. The interior design of the active lines is purposed to guide people to shuttle back and forth in different areas in an exploring way. The exaggerated sliding doors are designed on both sides of the aisle at the main entrance to divide SIN into two parts. The interior space is designed for division into multiple areas, which have stronger flexibility.

Stepping out of the elevator at the top floor, the textured wall surfaces on both sides of the aisle are applied to constantly reflect the light and create a crisscrossing effect. A 23m long snake-shaped pendant lamp, customized by 17，000 pieces of glass crystals, guides your view to the symbol at the entrance.

墙面均采用漆黑色调。

The wall applies the pitch-black color completely.

通过隔断使空间结构变得复杂。

In regard to the space arrangement, the space structure becomes complicated through the partitions.

红色营造一种心理上的酒吧氛围。

Red creates a psychological bar atmosphere.

菱形不锈钢片和天鹅绒拼接而成的饰面。

The veneer assembled by the diamond stainless steel pieces and velvets

各式花纹的椅子呼应伊甸园的主题。
The chairs in different patterns echo with the theme of Eden.
内部人流动线设计意在引导人们以探索的方式穿梭于不同区域。
The design of the internal streamlines of people intends to guide people to shuttle through the different areas in an exploring way.
五彩条纹的组合沙发绚丽多姿。
The colorful striped sofa set looks gorgeous.
吧台。
Bar Counter.

一个个小区域自成一体。

One after another small area becomes independent naturally.

楼上楼下的融会贯通。

The integration and connection between upstairs and downstairs.

花团锦簇，纷繁缭绕。

The flower clusters look bright and beautiful, but numerous and complicated.

柏丽雅女子会所 bailiya Women's Club

设计:林开新 / 面积:420 m² / 主要材料:大理石、墙纸、玻璃、软膜 / 坐落地点:福州 / 工程造价:60万元 / 完工时间:2009年3月 / 摄影:吴永长

Design:Lin Kaixin / Area:420 m² / Main Materials:marble、wallpaper、glass、pia mater / Location:Fuzhou / Project Cost:RMB 600 000 / Finish Time:Mar,2009 / Photography:Wu Yongchang

空间大量采用米白色软膜、皮质软包、纱幔等材料，并通过LED照明，尽显女性的健康与时尚。半开放顾问间设计为半包球式软膜结构，极具梦幻气息。设计选取欧式元素填充，使空间别有一番天地。软包墙面、化妆台等都彰显欧式简洁明亮的生活态度。在这个空间中，既能够体会到国际先锋的柔性改良，同时亦享受着符合自己审美理念的深层内核。在引进理念的同时，亦顺从大众审美，将二者完美地融为一体。

The space has greatly applied the materials such as off-white soft coat, soft leather package and voile curtain etc. together with the LED illumination to fully reveal the healthiness and fashion of the women. The semi-opening consulting room is designed as a hemispherical soft coating structure and has an extreme dreamy atmosphere. The design has selected the continental elements to fill the space with unique scenery. The soft packaged wall surface and dresser etc. have prominently shown the continental simple and bright living attitude. In this space, you both can experience the flexible improvement of the international pioneer, but also can enjoy the profound meanings matching your own aesthetic philosophy. It also complies with the public aesthetics while introducing the philosophy so that these two can be perfectly integrated together.

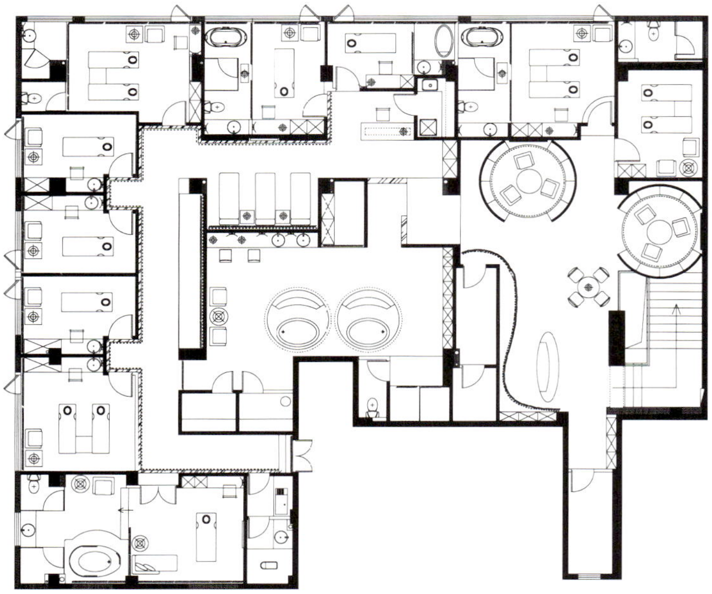

不同光影下的同一空间。
The same spaces under the different light and shadows.
半开放顾问间设计为半包球式软膜结构，极具梦幻气息。
The semi-opening consultation room is designed to a semi-ball type soft film structure of extreme dreamy atmosphere.

红色座椅分外抢眼。

The red chairs are extremely eye-catching.

白色皮质软包与米白色软膜相映成趣。

The white leather soft packs and off-white soft films set off with each other perfectly.

纱幔在灯光的衬托下尽显女性的健康与时尚。

The voile curtains have fully revealed the health and fashion of the females under the setoff of the light.

国中温泉 Guozhong Hot Spring

设计单位:南京广暄工程设计顾问有限公司 / 设计:陈广暄、马珺 / 面积:2800 m² / 主要材料:意大利雕刻白大理石、LG人造石、瓷质马赛克、古铜镜、皮质软包 / 坐落地点:南京 / 工程造价:840万元 / 完工时间:2009年3月 / 摄影:南京远景摄影设计有限公司

Design Unit:Nanjing Guangxuan Design / Design:Chen Guangxuan、Ma Jun / Area:2800 m² / Main Materials:marble、LGartificial stone、mosaic、bronze mirror、soft roll / Location:Nanjing / Project Cost: RMB 8 400 000 / Finish Time:Mar,2009 / Photography:Nanjing YuanJing photograph Design Co., Ltd.

如何让新古典风格与SPA水文化自然糅合在一起是本案所做的一次设计尝试。清澈、洁净、细腻、温和，是对空间风格的总体定位，同时也是设计师希望客人能通过空间行进次序而产生的心理感受。一个发光的玻璃盒子是建筑外观给人的第一印象，进入前厅两层各自带有图案的落地玻璃将轻盈外观印象延续到了室内，浅灰绿的色彩暗示空间中水文化的存在。纯白色的水区，由多种不同质感的白色材料组成。以对称为原则的设计方式，使整个空间带有鲜明的古典气质。服务区走廊的皮质软包墙面及包间电视背景墙面的粗纹绒布，给人以温存的触感。灰紫色的地毯形成一种四季皆宜的悦目感受。

How to naturally integrate the neoclassic style with SPA water culture is a design attempt of this case. Limpidity, Cleanness, exquisiteness and geniality are the general orientation to the space style. The designer also hopes that the customers can have a mental feeling through the moving sequence of the space. An illuminative glass box is the first impression of this exterior building on people. The two-layer floor glass with different figures at the entrance of the front hall has continued the simple and graceful visual impression to the interior. The light grayish green has hinted the existence of the water culture in the space. The pure white water area is composed of the white materials of different qualities. The design method in principle of symmetry has endowed the entire space with a vivid classic atmosphere. The leather soft packaged wall in the corridor of the service area and the coarse thread flannelette for the TV background wall in VIP rooms feel very warm and gentle. The grayish purple carpet has delivered a pleasing feeling suitable to all seasons.

带有图案的落地玻璃将轻盈外观印象延续到了室内。
The patterned floor glass has continued the lightsome visual impression to the indoor.
地面与墙面都采用花纹图案。
The floor and wall both apply the flower patterns.

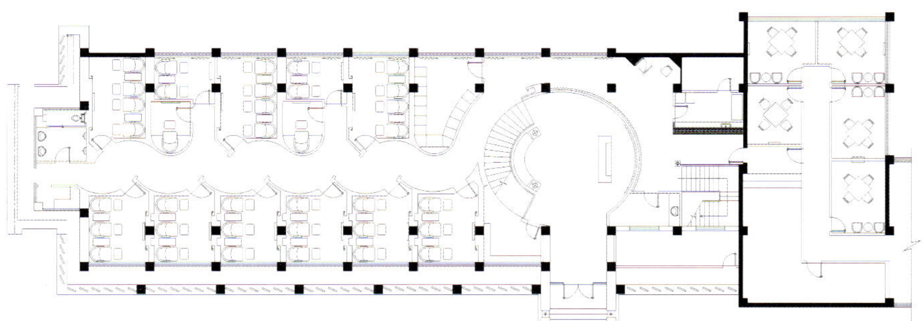

纯白色的水区，由多种不同质感的白色材料组成。
The pure white water zone is constituted by the various types of different white materials.
服务区走廊的皮质软包墙面。
The soft leather packed wall of the corridor in the service area.
粗纹绒布，给人以温存的触感。
The rough grained flannelette feels gentle.

若可生活时尚概念馆 Ruoke Life Fashion

设计:林开新 / 面积:250 m² / 主要材料:地板砖、软膜、金钢板、不锈钢 / 坐落地点:福州 / 工程造价:35万元 / 完工时间:2009年5月 / 摄影:吴永长

Design:Lin Kaixin / Area:250 m² / Main Materials:floor tiles、pia mater、stainless steel / Location:Fuzhou / Project Cost:RMB 35 000 / Finish Time:May,2009 / Photography:Wu Yongchang

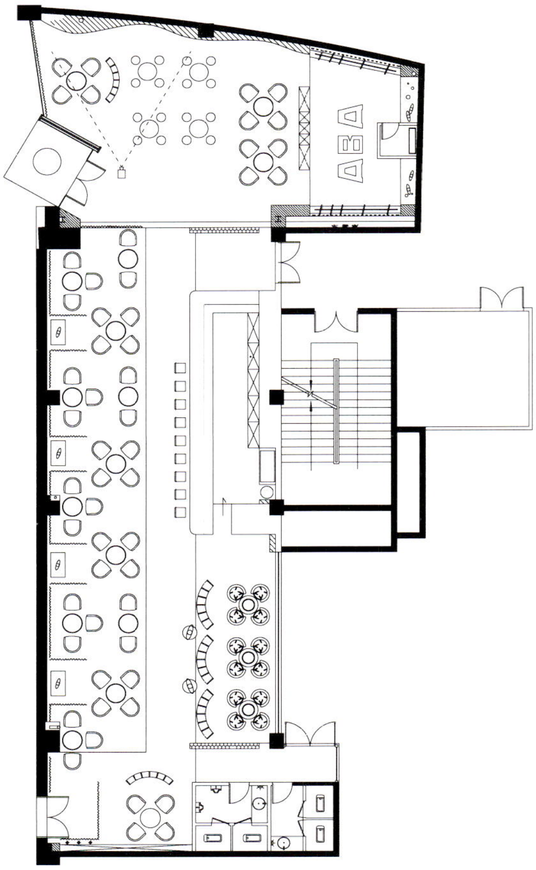

设计将流行、时尚、文化、感官的理念完美融合。时尚符号巧妙地运用到每一个环节，趣味和创意如舞动的精灵般吸引着人们的眼球。门口玻璃镶嵌的特大号纽扣、色彩变幻的发光吧台、橱窗边“守望”的模特、天花板上垂下的油画与镜面、造型独特的书架……当这些充满创意的装饰组合在一起时，那种时尚且灵动的感觉便奇妙地出现了。

The design has perfectly integrated the philosophies of fashion, style, culture and organs together. The fashion symbols are smartly applied to every link while the interest and innovation have attracted people's eyes just like the dancing fairies. The outsize buttons embedded on the glass at the entrance, the colorful luminescent bar counter, the "watching" models beside the show window, the oil paintings and mirrors hung from the ceiling, the book shelves in unique models …… when all these decorations are combined together, that fashionable and bright feeling will marvelously appear.

橱窗边“守望”的模特。
The "watching" model beside the shop window.
门口玻璃镶嵌着特大号纽扣。
The glass at the entrance is embedded with a extra big button.

照明指引了人们的视线。
The lighting directs the people' s vision.
色彩变幻的发光吧台。
The colorful versicolor bar counter.
天花板上垂下的油画与镜面。
The oil painting and mirror hanging down from the ceiling.

Yesbar Club Yesbar Club

设计单位:深圳市新冶组设计顾问有限公司 / 设计:陈武 / 面积:500 m² / 主要材料:石材、马赛克、地砖、实木复合地板、茶金镜、效果漆、乳胶漆、墙纸 / 坐落地点:重庆 / 工程造价:500万元 / 完工时间:2009年6月 / 摄影:钱翔

Design Unit:Newera Design / Design:Yellow Chen / Area:500 m² / Main Materials:stone、mosaic、floor tile、engineered flooring、special effect wallL paint、wallpaper / Location:Chongqin / Project Cost:RMB 5 000 000 / Finish Time:Jun,2009 / Photography:Qian xiang

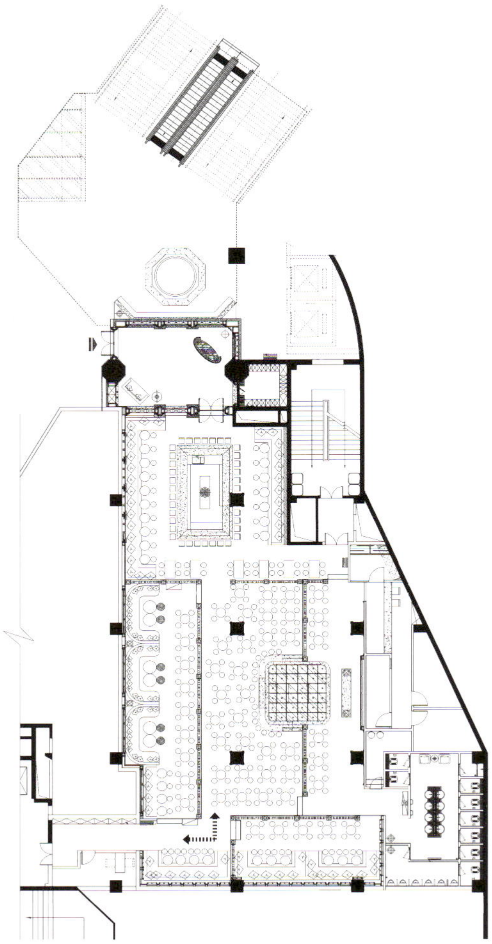

Yesbar Club 作为国际知名连锁品牌夜店，定位为高品位人士精致生活的时尚首选之地，打造城市娱乐的新地标。

设计师构建一个非主流现代阿拉伯宫殿式概念，异域奢华中尽显王者尊贵气派。以细节和品质的提炼和考究为本案的核心理念，务求以全新的方式去诠释人性深处的完美。

Yesbar Club as a international brand nightclubs is well-knowned, it' s the first choice of the high quality of life, and set entertainment city landmark .

Designers to build a idea, that is non-mainstream the concept of the modern Arab-style palace , it' s enough exotic and luxury to shows the distinguished king' s style! 500 ㎡ space cost $500 0000 yuan of investment costs, demonstrate the designers and owners of the common vision and approach, designer make sure that the core of the concept of the case is in pursuit of high quality, and use a whole new way to interpret the deep humanity Department perfect .

璀璨夺目的细部。
The resplendent details.
明暗有致的顶面。
The top with the proper light and shade.

斗猫 Boxing Cat Brewery

设计单位:BCA 装潢设计 / 设计:端木芸萱 / 参与设计:汪俊辉、夏芹、王唯鹿 / 面积:297 m² / 主要材料:紫铜、老木头、青砖、红砖、艺术涂料 / 坐落地点:上海 / 工程造价:40万元 / 完工时间:2009年7月

Design Unit:BCA / Design:Violet Tuanmu / Participatory Design:Wang Junhui、Xia qin、Wang Weilu / Area:297 m² / Main Materials:copper、old wood、blue brick、red brick、art coating / Location:Shanghai / Project Cost:RMB 400 000 / Finish Time:Jul,2009

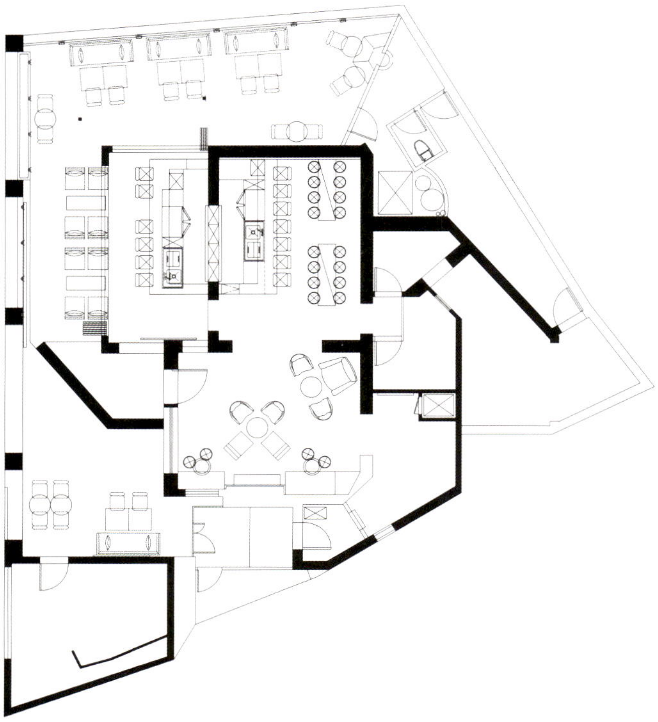

红色是斗猫设计的重头戏。无奈红色太艳丽，会让闷热的夏日灼烧起来，经讨论，选择褐紫红做为一楼的代表色,它包含了红色的热情、性感、自信与能量充沛，紫色的高贵与神秘，以及褐色系给人情绪安定、沉静、平和、亲切等意象。三楼也延续了红色，但这回却是鲜艳的唇膏红。二楼用餐区则是芥茉黄色的，透过艺术涂料创造斑驳岁月感觉。

除了色彩，嗅觉与触觉也是斗猫的设计重点。二楼，没有经过人工处理的原木错落有致，铺满整面墙，特有的气味就填满整个房间。

Red is a highlight of the cat-fighting design. However, as red is too f amboyant, it will burn the breathless summer. After discussion, maroon is selected as the representative color in the first floor. It contains the red passion, sex appeal, confidence and sufficient energy. The nobility and mystery of purple and the brow color series bring people with the feelings of stability, calmness, peacefulness, and friendliness etc. The third floor also continues the red, but this time it is the bright lipstick red. The dining area at the second floor is mustard yellow in order to create a time-flying feeling through the artistic coatings.

In addition to the colors, the smell and touch are also the highlights of the cat-fighting. At the second floor, the logs without manual processing are arranged orderly and cover the whole wall. Their unique flavors have filled up the whole room.

斗猫门头。
The cat-fighting overdoor.
吧台。
Bar Counter.

热情、性感的红色。

The ardent and sexy red.

原木错落有致。

The logs are strewn at random in order.

芥茉黄，透过艺术涂料创造斑驳岁月感觉。

The mustard yellow creates a feeling of mottled years through the artistic coatings.

露天休闲，古韵悠长。

Open and relaxing with profound antique charm.

红色是斗猫设计的重头戏。

Red is the highlight of the cat-fighting design.

延伸到室外的聚会定会非常浪漫。

The gathering extended to the outdoor must be very romantic.

东京秀客-意美姬造型 Tokyo Shower – Yimeiji Style

设计:高轶 / 面积:350 m² / 主要材料:玻璃钢、软包、线帘、高光混水漆 / 坐落地点:南京 / 工程造价:55万元 / 完工时间:2009年3月 / 摄影:朱子潺

Design:Gao yi / Area:350 m² / Main Materials:FRP、soft roll、cord thread、paint / Location:Naijing / Project Cost:RMB 550 000 / Finish Time:Mar,2009 / Photography:Zhu Zichan

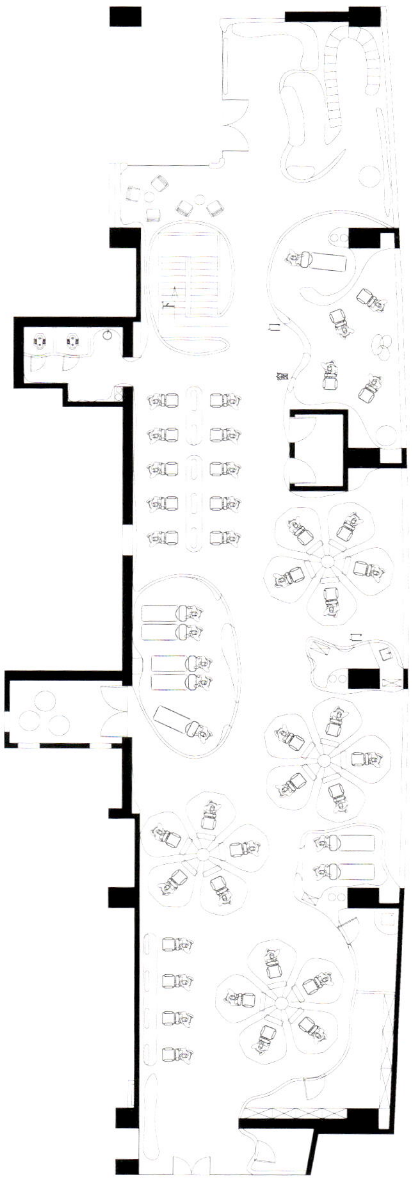

客人进门会被不规则的楼梯吸引。

The guests will be attracted by the irregular staircase once stepping in.

楼梯贯穿的上下空间。

The upper and lower spaces connected by the staircase.

秀客——意美姬是一家偏向日式简约风格的发廊。整体空间设计以白色为主，简洁中透着细致日式风格的主要特点，平面上根据其大空间比较"正"的特点，做了比较"整"的划分，让客人在空间感观上觉得更加通透。

客人进门会被不规则的壁饰和楼梯吸引，仿佛进入白色的未来空间，做一次在太空舱里美发的畅想。前台设计充满日式风格，镜面、楼梯及超大投影幕墙凸显现代感和梦幻色彩。

镜台采用花瓣式设计直通天花板，木制的大花瓣涂抹既美观又耐用，镜子的设计也呈现不规则的花瓣状，使空间活泼、跳跃。细节线帘的运用显示高雅品位，在光带的掩映下，褶褶生辉。

Shower – Yimeiji is a hair salon apt to Japanese–type simple style. The whole space is designed mostly with white and mainly characterized as a fine Japanese style in the simplicity. Its plan is divided quite "integral" as most of its space is "square" so that the customers visually feel its space more transparent.

The customers will be attracted by the irregular wall ornamentation and staircase once entering the door, as if they have entered a white future space and have a free thought of hairdressing in the space capsule. The reception area is designed full of Japanese style. The mirror faces, staircase and super large projection curtain wall have prominently shown the modern sense and dreamy colors. The dressing table is designed in a petal style directly up to the ceiling. The large wooden petals are both beautiful and endurable. The mirrors are also designed in an irregular petal style to make the space active and jumpy. The detailed string curtains have shown elegance and taste, and look splendid by the setoff of the light strips.

IMAGE
意美姬

细节线帘的运用显示高雅品位。
The application of the detailed yarn curtain shows the elegant taste.
镜子的设计呈现不规则的花瓣状。
The mirror is designed in an irregular petal form.
镜台采用花瓣式设计直通天花板。
The dressing table has applied the petal design directly up to the ceiling.

白色的未来空间，不妨做一次在太空舱里美发的畅想。

The white future space provides a free hairdressing thought in the space capsule.

华侨豪生SPA会所 Howard Johnson SPA Club

设计单位:HBS+宁波红宝石装饰设计有限公司 / 设计:宋国梁 / 参与设计:郑海雁 / 面积:2000 m² / 主要材料:木皮、大理石、墙纸、实木板、花岗石毛胚 / 坐落地点:宁波 / 工程造价:400万元 / 完工时间:2008年10月 / 摄影:刘鹰

Design Unit:HBS Decorative Design Co., Ltd. / Design:Song Guoliang / Participatory Design:Zheng Haiyan / Area:2000 m² / Main Materials:veneer、marble、wallpaper、solid tread、granite / Location:Ningbo / Project Cost:RMB 4 000 000 / Finish Time:Oct,2008 / Photography:Liu ying

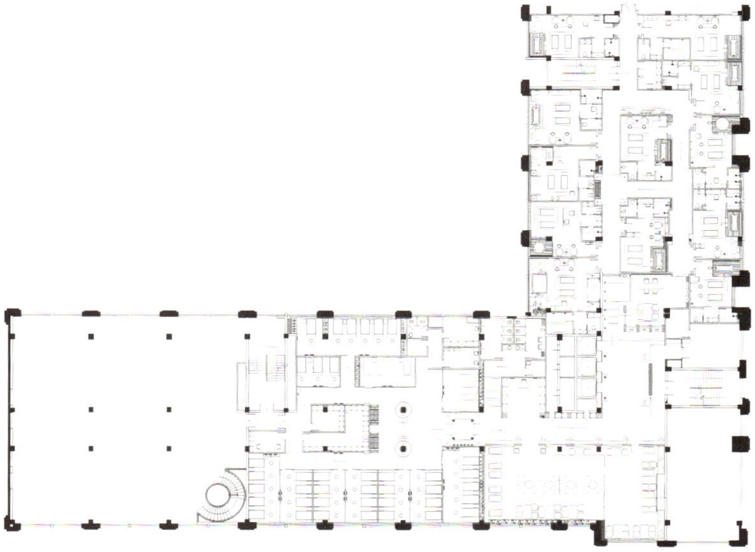

本案原是坐落在酒店的写字楼办公层，原有的机构和设备条件并不理想，过低的层高也给设计师带来不少挑战，最后用视觉错位手法弥补。

会所里的文化艺术中心和长廊，成为体验巴厘文化的中心纽带。以巴厘岛著名艺术重镇乌布命名的乌布工坊可以售卖它本土的手工艺术品，不出国门就可以选购巴厘木雕、铜艺手工艺品。SPA以水为核心，从接待区开始到每间包房，点点烛光倒映水上。包房设计风格各异，毛石、竹木地板、天然木材的运用体现以自然、健康为主题的理念，并从听、视、味、触、嗅五感体现SPA给人带来的立体感受。为体现原汁原味，所有饰品和关键材料都是远赴巴厘岛精心挑选定制的。

This case is located in the office floor of the office building in the hotel. The original organization and equipment facilities are not very ideal. The over-low floor height also brings great challenges to the designer and is finally offset by the visual staggering way.

The culture and art center and long corridor in the club has become the central link of experiencing Bali culture; Ubud artwork workshop named after Bali' s famous artwork town – Ubud can sell its local handicrafts. You can buy Bali wooden carvings and copper handicrafts without going abroad; SPA is centralized in water. From the reception area to every VIP room, dots of candle light are reflected on the water. The design styles of each VIP room are various while the application of the bamboos, bamboo floor and natural wood materials have shown its philosophy of naturalness and healthiness and reflected the three-dimensional feelings from hearing, sight, taste, touch and smell brought by SPA. To demonstrate its originality, all furnishings and key materials are delicately selected and customized from Bali.

从家具到配饰，处处体现巴厘风情。

From the furniture to the ornaments, everywhere is revealed with Bali charms.

深邃神秘的过道。

The deep and mysterious corridor.

长廊，成为体验巴厘文化的中心纽带。

The long corridor becomes the central link of experiencing Bali culture.

点点烛光。

The dotted candle light.

天然木材的运用体现以自然、健康为主题的理念。

The application of the natural timber has reflected the philosophy of the natural and healthy theme.

包房设计风格各异。
The design styles of VIP rooms are different from each other.
毛石的运用返璞归真。
The application of the rubbles seems returning to the nature.

荷欢印巴风情餐吧 Hehuan

设计单位:成都优佰文化传播有限公司 / 设计:孙亮 / 面积:420 m² / 主要材料:实木、仿古砖、青石砖、肌理涂料、布艺 / 坐落地点:成都 / 工程造价:60万元 / 完工时间:2008年6月 / 摄影:成都视觉空间商业摄影机构

Design Unit:You Bai Culture Spreading Co., Ltd. / Design:Sun liang / Area:420 m² / Main Materials:solid wood、rustic tile、brick、coating、cloth / Location:Chengdu / Project Cost:RMB 600 000 / Finish Time: Jun,2008 / Photography:VISION SPACE

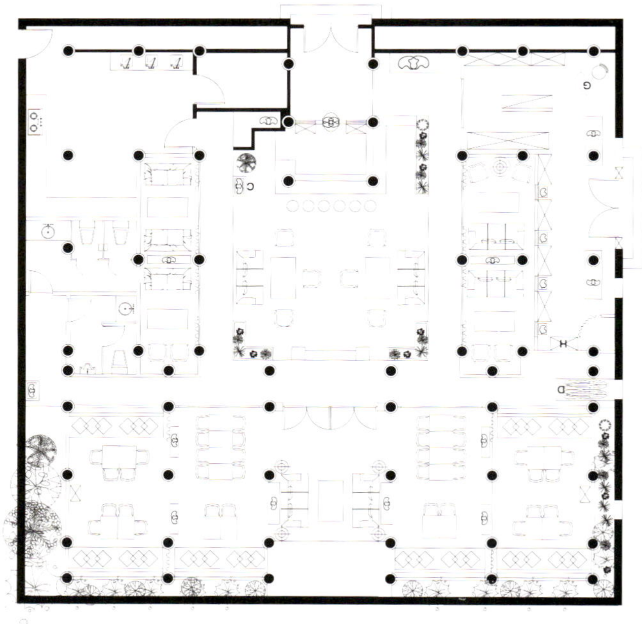

荷欢——一个极其暧昧的名字，如何通过具像的“形”来表达抽象的意识形态，设计的重点就在于此。于是，柔软、温暖、慵懒、淳朴，就成了关键词。

设计师对原有的仿古建筑做了修改，在不改变主体风格的情况下，在建筑立面上使用了玻璃，丰富了室内环境的光线；在室外的吧台上新建了屋檐以压低空间，增加舒适感和安全感。整个空间所使用的材料大都以实木、肌理涂料、仿古砖为主，以还原建筑本来的风貌。在完成了主体改造后，最重要的就是体现“印巴风情”的特点了。软饰在此起到了关键作用，色彩浓重的纱幔、台布，木雕的屏风，竹编的窗帘，再到家具、陈列品的选用，都围绕这个主题。在特色灯具的映衬下，整个环境变得朦胧、暧昧。红花、绿树、暖阳，如果再来杯特产的whisky，你已身在异域……

Hehuan: an ambiguous name, in accordance with this idea, we focus on how to use the exact "shapes" to express the nonfigurative ideology. Therefore, the key words here are: softness, warm, sluggishness and nature

Designer made some changes for the pseudo-classic building; Like putting glass on the outside wall to brighten the inside without changing the style of the building. Putting eaves upon the bar counter to lower the space and expand the feeling of comfortableness and safeness. Using the natural materials like wood, texture paint and glazed tile to restore the original looking of the building… After the building's renovation, another important factor—decorations brings out the tropical character of Indonesia and Bali Island, you will see the colorful muslin, fabric and bamboo curtains, typical wooden screen, even for the furniture and the decorative arts all are emphasizing the theme; Illuminated by the special lights, you will be able to experience the ambiguous and exotic feelings with the red flower, green tree and the sunny sunshine… … cheers!

光影世界里的年代感。

The sense of years in a world of light and shadow.

露天餐桌别有一番异域风情。

The open dining table has a unique foreign charm.

空间使用的材料大都以实木、肌理涂料、仿古砖为主。

The materials applied in this space are mostly the wood, textured coatings and antique bricks.

软饰在此起到了关键作用。

The soft decorations also take a key function.

家具、陈列品的选用，都围绕印巴风情的主题。

The selection of the furniture and displays are centered on the theme of Indian and Pakistan charms.

王春美容院CFC摩登馆

CFC Modern Club of Wangchun Beauty Salon Beauty Salon

设计单位:南京陆文星室内设计事务所 / 设计:陆文星 / 面积:1100 m² / 主要材料:黑金沙石材、毛石、泰国织物、镜面马赛克、麦格丽影木、金格墙纸 / 坐落地点:南京 / 工程造价:600万元 / 完工时间:2009年2月 / 摄影:陆文星

Design Unit:Nanjing Lu Wen Xing Design / Design:Lu Wenxing / Area:1100 m² / Main Materials:stone、rubble、cloth、mosaic、wood、wallpaper / Location:Nanjing / Project Cost:RMB 6 000 000 / Finish Time: Feb,2009 / Photography:Lu Wenxing

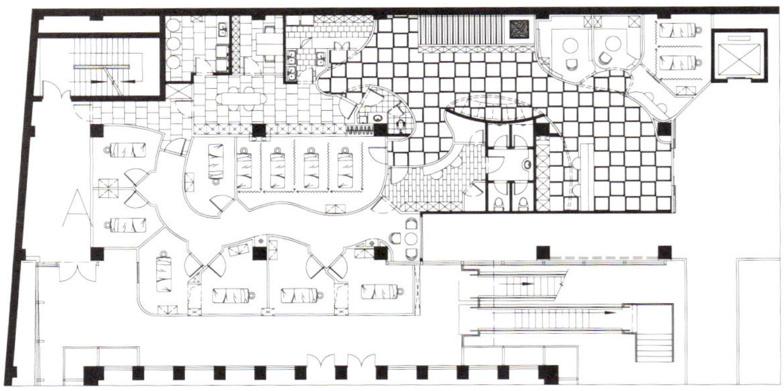

王春美容院对光的要求很高，除了运用专业照明灯具，还采用光纤灯展现流光溢彩的魅惑，提升了空间的梦幻质感，诠释出迷人的静态光源美。缀花玻璃与幽蓝冷光和此处的暖色形成对比，冷暖色颠覆着神经，丰富了空间的格调。佛龛饰品象征着某种地域风情，带给客人尊贵而神圣的空间体验，它突出气氛的独特性、描绘着典雅庄重而又安宁静谧的视觉。经典材质搭配以及精致饰品，营造出美疗空间的摩登意境。

Wangchun beauty salon has high requirements for the light. In addition to the application of the professional lighting fixtures, the optical fiber lamps are also applied to show the brilliant seductive charms, upgrade the illusionary sense of the space and deliver a charming beauty of quiet light sources. The flower-decorated glass and secret blue cold light have formed a contrast with the warm colors here. The cold and warm colors are overthrowing the nerves and have enriched the space style. The niche for a Buddhist statue has symbolized the certain regional customs and conditions and brought the people the noble and sacred space experience. It has outstood the uniqueness of the atmosphere and described the elegant, grand, peaceful and quiet vision. The classic material arrangement and the delicate furnishings have created a modern atmosphere of a beauty space.

楼梯区两侧毛石在灯光的配合下，极具肌理感。

The rubbles on both sides of the staircase area look extremely textured under the light.

惟妙惟肖的佛龛饰品。

The remarkably lifelike niche for a Buddhist statue.

光纤灯展现流光溢彩的魅惑。

The optical fiber lamp is demonstrating a gorgeous charm.

七彩斑斓的光影世界。
A colorful world of light and shadow.
迷人的静态光源美。
The charming quiet light beauty.

BEAUTY SALON

镜面马赛克铺置的大厅。
The mirror mosaic covered lobby.
公共区马赛克抢占了墙面。
The mosaic occupies the wall in the public area.
艳丽热烈的火红夺人眼球。
The colorful passionate flaming red attracts your eyes.
咨询区墙面的荷花纹案绚丽多姿。
The lotus patterns on the wall in the consultation area are gorgeous and graceful.

新锦秀国际娱乐会所 Xinjinxiu International Entertainment Center

设计单位:杭州金白水清悦酒店设计工程有限公司 / 设计:张丰义、王建强 / 参与设计:陈伟强、曹龙、孟宇铮 面积:5600 m² / 主要材料:淋漆板、大理石、玫瑰金不锈钢、金镜、绒布 / 坐落地点:杭州 / 工程造价:3000万元 / 完工时间:2009年5月 / 摄影:陈乙

Design Unit:Hangzhou JinBaiShuiQingYue Hotel Design .,LTD. / Design:Zhang Fengyi、Wang Jianqiang / Participatory Design:Chen Weiqiang、Cao long、Meng Yuzheng / Area:5600 m² / Main Materials:marble、rose gold stainless steel、gold mirror、flannel / Location:Hangzhou / Project Cost:RMB 30 000 000 / Finish Time:May,2009 / Photography:Chen yi

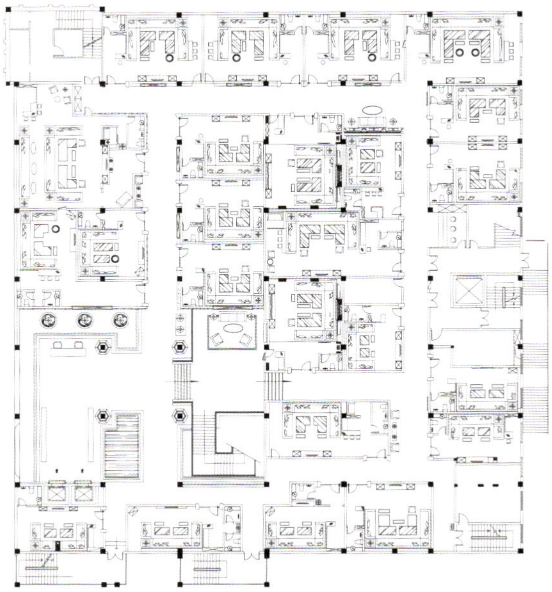

新锦秀国际是杭州新开的一家顶级奢华娱乐会所。无论从体块、造价，还是材质、配饰等，都极尽恢宏之势。尊贵的金色笼罩着大堂，天花、扶手，甚至地面，都像被镀上一层浓烈的彩妆。公共休息区热烈鲜艳的红色镶金边沙发与整个环境十分协调。递进式流线型天花的设计也在一定程度上丰富了顶面层次。包间背后的墙面成了展示精品的陈列架，每件作品都是精挑细选，以求匹配奢华品味。

Newly established in Huangzhou, China, Xinjinxiu International Entertainment Center is top-grade entertainment center. The entertainment center is full of grandeur in the aspects of size, construction costs, materials, and fittings. The hall is decorated by the golden color (the color only for royal families in ancient China), which shows respects for guests. It seems that vivid fresh colors are everywhere, including the ceiling, armchairs, or even the ground. The gilt-edged bright red sofa in the public lobby is harmonious with the whole environment. To a certain extent, the design of the stepwise and streamline ceiling patterns makes the ceiling more changeable. Compartments can receive numerous visitors. Excellent works are displayed on the back walls, and each piece is carefully chosen to match your luxurious taste.

一、二、三层的前进楼梯。
Forward stairs in Floor 1, 2 and 3.
包间背景墙俨然小型精品陈列架。
Background wall of the compartment look like a small size shelf for elaborative articles.

紫色笼罩的小喷泉。
Purple small fountain.
精雕细琢的镂空隔断。
Finely carved hollow partition walls.
热烈艳丽的红色沙发分外惹人注目。
Fire-like red sofa attracts much attention.
递进式流线型天花丰富了顶面层次。
Progressive streamline ceiling enriches the layers on the top.

丽歌量贩KTV Lige KTV

设计单位:杭州金白水清悦酒店设计有限公司 / 设计:张丰义、王建强 / 参与设计:陈伟强、曹龙、孟宇铮 / 面积:3200 m² / 主要材料:大理石、UV板、灰镜、微晶石 / 坐落地点:宁波 / 工程造价:1000万元 / 完工时间:2008年7月 / 摄影:刘鹰

Design Unit:Hangzhou JinBaiShuiQingYue Hotel Design .,LTD. / Design:Zhang Fengyi、Wang Jianqiang / Participatory Design:Chen Weiqiang、Cao long、Meng Yuzheng / Area:3200 m² / Main Materials:marble、UV board 、grizzly mirror、microlite / Location:Ningbo / Project Cost:RMB 10 000 000 / Finish Time: Jul,2008 / Photography:Liu ying

丽歌量贩KTV处于东部新城的中心区域，独立楼宇结构。为求达到以客为尊的目的，色调、材质和造型设计都力求给宾客舒适的享受。各类纯量贩式包房，每间房都具有特色，独树一帜。是时尚年轻一族的娱乐新地标。

Located in the central part of the new eastern district, Lige KTV a structurally independent building and is always customer-oriented. All staff of Lige KTV make every effort to bring comfort and satisfaction to customers in the following details: colors, materials and styling design. All compartments show unique characteristics, serving as a new choice of entertainment for fashionable young people.

夸张的大幅喷画另类、张扬。

Exaggerated large paintings in unique and grand styles.

入口接待台绚丽多姿。
Colorful reception desk at the entrance.
同样的设计灵感由过道变到顶面。
A same thought of design connects the corridor to the ceiling.
树叶瓣镂空屏风经典与华丽并重。
Leaves-like hollow screen integrated with classicism and luxury.

普拉那啤酒屋 Paulaner Brauhaus

主要材料:木材、红砖、画框、乳胶漆 / 坐落地点:苏州 / 摄影:潘宇峰

Main Materials:wood、brick red、frame、emulsion paint / Location: Suzhou / Photography:Pan Yufeng

普拉那是德国著名的啤酒品牌。空间设计希望营造欢快活泼、随意自然、年轻雅皮的气氛。裸露的红砖、自然的木材、硕大的啤酒容器完全展现在人们眼里，减弱了人与建筑、与空间的陌生感。地面和顶面都使用原汁原味的木材，仿佛地中海吹来的一室天然。静坐其中品味着现酿的啤酒，清醇在口，怡然在心。

Paulaner is a famous beer brand in Germany. The purpose of the space design is to create a happy, lively and natural atmosphere favored by young yuppies. Uncovered bricks, natural wood, large beer containers, all these visible objects make you more familiar with the buildings and the space. Constructed by natural wood materials, the ground and ceiling make you feel the smell of the nature from the Mediterranean. Just sit still to enjoy a cup of fresh beer. A pure taste will bring you satisfaction and happiness.

开放式楼梯。
Open stairs.
地面和顶面都使用原汁原味的木材。
Purely original wooden materials for the floor and ceiling.
对称的光带丰富了平淡无奇的顶部。
Symmetrical light belts enriches the plain ceiling.

1F

原有建筑梁柱在其中丝毫不觉突兀。

The original beams and columns of the building do not look sudden at all among it.

餐厅。

The dining room.

电视背柜与木框镜面形成一个开启的动态关系。

The back cabinet of television has formed an opening dynamic relationship with the wood framed mirror.

看得见风景的房间。

The room where the scenery can be seen.

卫生间。

The washroom.

老电话。

The antique telephone.

完美生活 Perfect Life

设计单位:哈尔滨唯美源装饰设计有限公司 / 设计:张奇永 / 面积:650 m² / 主要材料:石材、玻璃、涂料、瓷砖
坐落地点:齐齐哈尔 / 工程造价:300万元 / 完工时间:2008年12月 / 摄影:张奇永

Design Unit:WeiMeiYuan Design / Design:Zhang Qiyong / Area:650 m² / Main Materials:stone、glass、coating、ceramic tile / Location:Qiqihar / Project Cost:RMB 3 000 000 / Finish Time:Dec,2008 / Photography:Zhang Qiyong

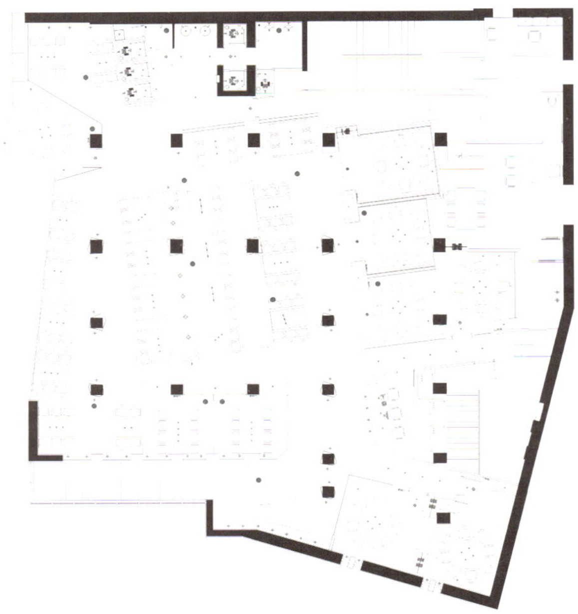

透过玻璃依稀可见老建筑的柱子。
The columns of the old building can be seen vaguely through the glass.
印有图案的玻璃隔断。
The patterned glass partition.

站在这个需要重新设计并改造、装修的房子里，面对不甚合理的格局和较为零散的空间时，设计师瞬间跳出“完美”这个字眼。身处一个空间，设计师最应该做的是整合。先规划好布局，再摆入空间所需的实用性物品，最后进行打扫与清洁，让空间实用性与功能性兼备，美感与舒适并存。明亮的落地窗、大片的阳光、新鲜的空气和水以及绿色，这些都是食物的绝好搭配。身处这样的环境中，人们必会在追寻完美的路途上畅快前行。

Standing in this house demanding the re–design, modification and re–decoration and confronted with the unreasonable layout and quite scattered spaces, the designer instantly jumped out of this superficial wording ‘perfection’. Seated in a space, what a designer shall do most is re–arrangement. It shall be planned, re–arranged and placed with the practical articles necessary to the space before the final cleaning so that the space can both have the practicality and functionality as well as the co–existence of aesthetics and comfortableness. The bright French Windows, large area of sun light, fresh area and water as well as the greenness are absolutely the best match for the foods. Involved in such an environment, You will certainly move forward smoothly in the journey of pursuing to the perfectness.

洗手间。
The washroom.
时尚纹样。
The fashionable patterns.
背光。
Against the light.
实木板在顶部的造型。
The model of the solid wood boards on the top.

贝斯酒吧 Bass Bar

设计单位:宁波市高得装饰设计有限公司 / 设计:范江 / 参与设计:丁伟哲 / 面积:200 m² / 主要材料:松木、大理石、水泥、海藻泥涂料 / 坐落地点:宁波 / 工程造价:20万元 / 完工时间:2008年6月 / 摄影:文宗博、潘宇峰

Design Unit:Gotel Design Studio Limited. / Design:Fan jiang / Participatory Design:Ding Weizhe / Area:200 m² / Main Materials:pine wood 、marble、cement、coating / Location:Ningbo / Project Cost:RMB 200 000 / Finish Time:Jun,2008 / Photography:Wen Zongbo 、Pan Yufeng

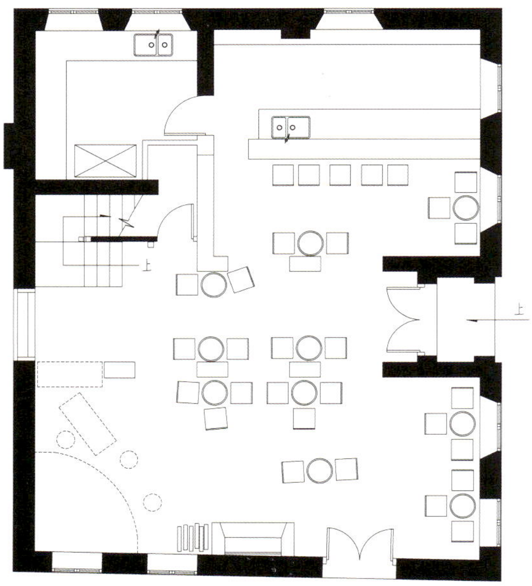

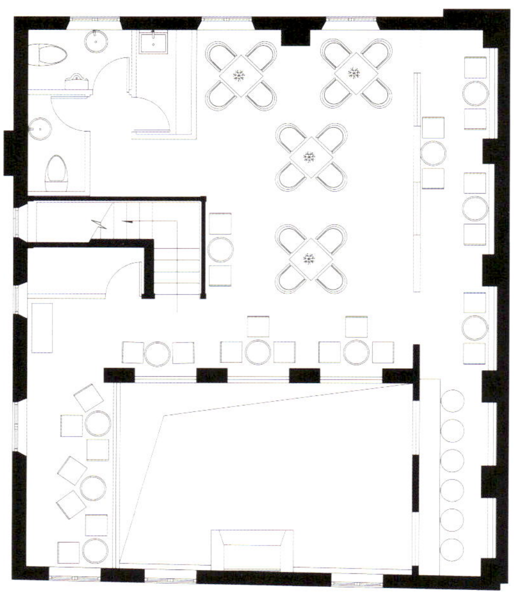

太古洋行建筑特征明显，业主的经营思路从咖啡厅、餐饮最后跳到酒吧，希冀是一个飘着爵士乐、有酒有茶的休闲场所。二楼楼板被拆除一部分，加入有拱形门洞的实墙，上下贯通。一层以水泥地坪为主，吧台、壁炉处、进门处是大理石马赛克拼花，铺设时故意略带残缺，边缘与水泥地坪进行自然过渡。二层保留原先的老地板。阳台用玻璃围了进来，补充欠大的营业空间。

家具基本定做，高脚褐色皮椅面的吧椅，圆形松木桌子，符合空间特点也是出于经济考虑，卫生间、酒柜上部是银镜正反两面刻花，刻的是以前欧洲人诙谐有趣的生活场景，以剪影的形式艺术地再现。墙上挂着老上海的月份牌，时尚、摩登、开放的气息扑面而来。二层的一隅放上了高背草绿丝绒皮椅，业主收藏的大喇叭唱机放上黑木胶片，胶片轻轻旋转，那歌声把我们带到那个久远的年代。

The Swire Pacific building has clear characteristics. The business operation philosophy of the owner jumped from the cafe, restaurant and finally to the bar room. The floor slab of the second floor has been partially removed and added with a solid wall with an arch door opening so that the upstairs and downstairs are connected together. The first floor is mainly the cement flooring while the bar counter, fireplace and entrance are covered with the marble mosaic figures. The incompleteness is deliberately left during the paving. The second floor has reserved the original old floor boards while the balcony is enclosed by the glass to make up the insufficiently big operation space.

Most of the furniture is customized to answer to the space characteristics and to make economic consideration. The toilet and the upper of the wine cabinet are the silver mirror engraved with flowers on both sides and are artistically revealed in a form of silhouette. The calendar art of the old Shanghai are hung on the wall and you are greeted with the fashionable, modern and opening atmosphere. One corner of the second floor is placed with a high-end grass green velour leather chair. The big horn phonograph collected by the owner is placed with a black record. The record gently rotates and the singing brings us to that distant time.

酒吧外观。
The bar appearance.
加入拱形门洞的红砖实墙，上下贯通。
The red brick wall with the arched door is linking upstairs and downstairs.

壁炉。

The fireplace.

挑空的阁楼上高凳有更多观望的视角。

There are more watching angles while seated on the high stools in the overhanging attic.

存在老留声机仿佛存在老的音乐。

The existing old gramophone seems like the existence of old music.

木格上方银镜正反刻有以前欧洲人生活场景的图案。

Both sides of the silver mirrors on the wood grids are carved with the patterns of the European living scenes in the old times.

新加墙体与原有木架梁的关系。

The relationship between the newly added wall and the existing wood beams.

门拱形成的区域空间。

The space formed by the door arches.

诺富特酒店 Novotel Hotel

设计单位:法国纳索设计工作室 / 设计:Margaux Lhermitte、Adam Fang / 面积:16 000 m^2 / 主要材料:木材、大理石、地毯 / 坐落地点:香港 / 完工时间:2008年5月 / 摄影:申强

Design Unit:NACO Architectures / Design:Margaux Lhermitte、Adam Fang / Area:16 000 m^2 / Main Materials:wood、marble、carpet / Location:Hong Kong / Finish Time:May,2008 / Photography:Shen qiang

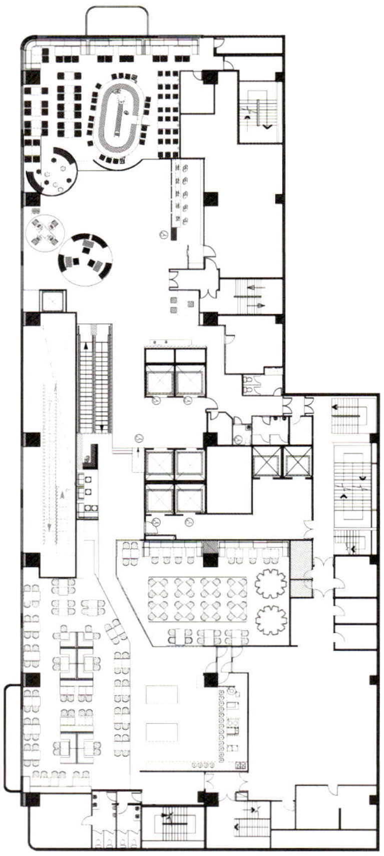

项目是一个酒店的翻新设计。建筑本身的条件较差，大厅和入口分别被修建在不同的楼层。客人到达酒店后不能直观了解大堂全貌。为此设计了一个通道，大理石地砖和天花板木材的纹路相呼应，引领客人乘坐扶手电梯通往上层接待区。接待室后部的设计特点在于吊顶的高度，弧形设计给整个空间注入温暖、圆滑的感觉。

酒吧朝向大厅，但进入其中却发现是一个私密场所。弧形玻璃墙起到隔音作用，加之柜台前的爵士乐，给人一种深邃迷离的感觉。餐厅风格独特，中间是可视厨房和自助餐区。大理石设计既简洁又独特。错落排列的瓷砖使整个餐厅古色古香。客房区灰色条形石头和镜子的完美结合，柔软且安静。顶层是高级休闲区，两扇Z字形的滑动门，使人印象深刻。

It is a renovation design for a hotel. Its own conditions are quite bad. The lobby and the entrance were built separately at the different stories. The guests can not visually know the full view of the lobby after entering the hotel. Due to this reason, a passage is designed, where the marble floor tiles echo with the texture of the wood ceiling to lead the guests to take the escalator to the upstairs reception area. The design of the rear of the reception room is featured at the ceiling height. The arc design has infused the whole space with a warm and smooth feeling.

The bar is facing towards the lobby while it is a intimate place once entering. The arc glass wall has functioned as sound insulation. The restaurant has an unique style, in the middle of which is the visible kitchen and buffet area. The marble design is both simple and special. The whole restaurant looks antique due to the randomly strewn ceramic tiles. The perfect combination of the gray strip stones and mirror in the guest room area looks soft and peaceful. The top is the superior recreation area and bivalve of Z-shape sliding doors are very impressive.

总服务台。

The general service desk.

餐厅。

The restaurant.

电梯通道。
The corridor to the lift.
水吧台。
Drink bar counter.
时尚纹样。
Fashionable patterns.

线 · 魅力 Line·Charm

设计:林开新 / 面积:99 m² / 主要材料:PVC地板、墙纸、木面板、亚克力板 / 坐落地点:福州 / 工程造价:8万元 完工时间:2008年6月 / 摄影:吴永长

Design:Lin Kaixin / Area:99 m² / Main Materials:PVCfloor、wallpaper、wooden panel、acrylic board / Location:Fuzhou / Project Cost:RMB 80 000 / Finish Time:Jun,2008 / Photography:Wu Yongchang

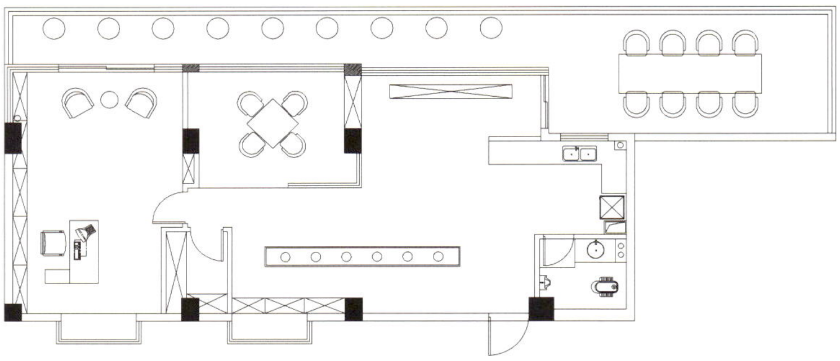

空间原是两室一厅的住宅用房。保留原住宅的厨房和卫生间位置并对厨房进行扩大和改造，成为展示空间里的接待吧台。原住宅的主卧改为经理办公室，次卧改为接待区。线是本案设计元素的主角，组织成列、成面、成格子。吊顶上的线或面利于专业照明灯具的安装和过多管线的预埋，组织出严谨的美感；展示架的线成列、成格子严谨且有趣味，并达到展示功能的需要。在洽谈区和展区之间的隔断上，线被有意识的抽开一些，不仅开辟了展示空间，同时在LED变色光盒衬托下，给空间多了一份梦幻的体验。线设计元素的成型是白色的，空间基础底色是暗灰的，这种明度的强烈对比诠释了空间的时尚含义。

The space used to be a residence with two bedrooms and one living room. The positions of the original kitchen and washroom are reserved while the kitchen has been also expanded and modified into a reception bar counter in this exhibition space. The original master' s bedroom is changed to the manager office while the secondary bedroom is changed to the reception area. The line is the hero in our design elements to form into the lines, surfaces and panes. The lined surface in the ceiling is favorable for the installation of professional lighting fixtures and embedment of excessive pipe lines. In this way, the compact aesthetic can be organized; the lined rows and panes in the exhibition shelves are compact and interesting and can met the requirements of exhibition functions. In regard to the partition between the discussion area and the exhibition area, some lines are intentionally taken out, which not only has developed the exhibition space, but also adds a dreamy experience to the space under the setoff of the LED color-changing light box. The completed shapes of the line design elements are white while the background color of the space is dark gray. Such strong contrast on the brightness has interpreted the fashionable meaning of the space.

接待台。
The reception desk.
展示架形成的线形。
The linear shape formed by the display shelves.

过道。
The corridor.
洽谈区。
The discussion area.
前厅。
The front hall.

中森名菜 Nakamoni Akina

设计单位:文格空间设计顾问（深圳）公司 / 设计:林文格 / 面积:16 000 m^2 / 主要材料:锈石、毛面大理石、埃比蒙木、墙纸、镜钢、艺术玻璃、澳洲砂岩 / 坐落地点:深圳 / 工程造价:1500万元 / 完工时间:2008年11月 / 摄影:吴永长

Design Unit:Wen Ge Space Design Adviser (Shenzhen) Company / Design:Lin Wenge / Area:16 000 m^2 / Main Materials:stone、marble、wood、wallpaper、steel、art glass、Australia sandstone / Location: Shenzhen / Project Cost:RMB 15 000 000 / Finish Time: Nov,2008 / Photography:Wu Yongchang

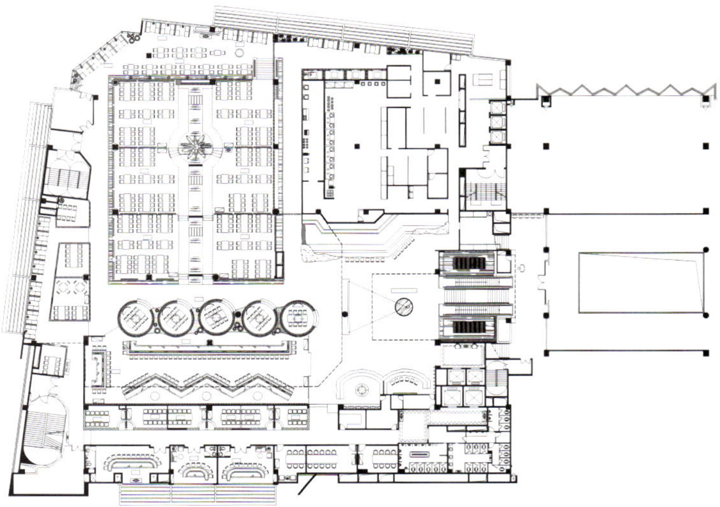

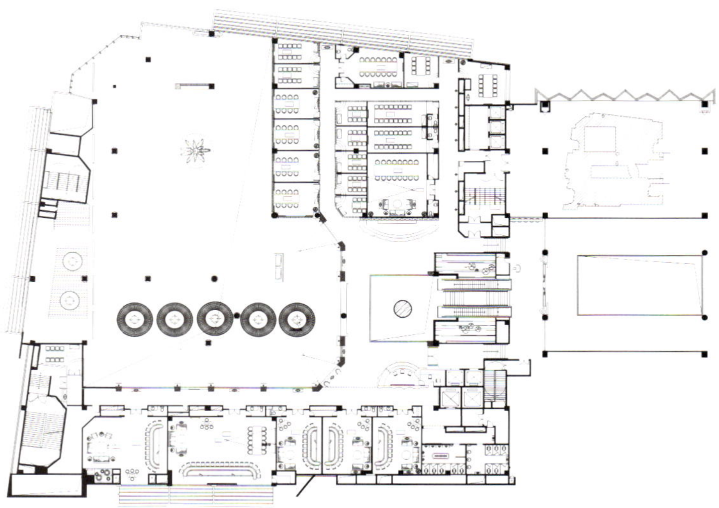

此案定位为现代和风。选取樱花、清酒、纸灯笼等最具代表性的直观印象符号，演化为装置艺术品；采用原始古朴材料，如毛胚石材、砖、木等；借鉴日本传统造景技艺以及对面积利用最大化的考量方式。通过夸张、变形、简化、模拟等多样化的处理手法，将古典日式元素融入时尚商业空间。水幕循环系统大幅度减弱室内造景难度，园林流水源源不绝，增添灵动氛围。

原始建筑形态特殊，平面中有一块突兀的抬高区域，顶部是巨大的透光中庭。设计师巧妙地将劣势转化为优势，在突起部分营造人工园林，形成跌落水景观。无数灯珠构筑成一树樱花，立于水池中，成为空间焦点。设计师充分利用过渡区域，在大厅散座和铁板烧烤台之间加设五圈雅座。雅座选取鸟笼这种东方意象，扩大并简化为半围合空间，在开放和私隐之间取得平衡。五组大圆套小圆的造型序列排开，视觉冲击强烈，无形中掩映了两根不规则立柱，同时实现区域间的自然过渡，再一次化解空间难题。

中庭层高充裕，增设3000 m^2夹层。就餐模式进一步细分，面向中庭一侧的包房，采用玻璃栏围合的半开放式格局，完整展现大厅景观。另一侧则是全封闭式包房，满足尊贵私密的消费心理。

通过对整个餐饮空间的解构，形成散座区、雅间区、铁板烧烤区、榻榻米包间、隐私包间等五种就餐模式，提供全开放、半开放、全封闭三种就餐环境，配合相应收费标准，人为创造出极具市场吸引力的超大空间。

This case is orientated at the modern Japanese style. The most typical visual impressive symbols such as cherry blossoms, Japanese wine and paper lanterns are selected as the decorative artworks; the primitive antique and simple materials such as blank stone materials, bricks and woods; the traditional Japanese gardening techniques as well as the consideration of maximizing the area are borrowed for this design. The recycling system of water curtain has greatly reduced the gardening difficulties while the continuously running water has added a sapiential atmosphere.

The original building form is special. There is a suddenly raised area in the plane while the huge photic atrium is on the top. The designer has built an artificial garden in the protruded area, which forms into a water-falling sight. One cherry tree standing in the pool becomes the focus. The private rooms are simplified into the semi-enclosed spaces and have achieved the balance between the opening and privacy. The five sets of the models that big circle surrounds small circles are arranged in sequence to realize the natural transition among the areas and further settle the space problem again.

The floor height in the atrium is enough to arrange a 3000 m^2 mezzanine. The dining pattern is further subdivided and the semi-opening layout has integrally displayed the sight in the hall. On the other side are the entirely closed VIP rooms to meet the intimate consumption psychology.

The dining space is divided into five dining patterns, providing three types of dining environments with appropriate charging standards. A super large space of marketing attractions is created artificially.

清酒架。

Japanese wine rack.

灯笼状的形态。

The lantern-like form.

体量巨大的布灯。
The giant clothing lamp.
料理台。
The food table.
圆形木格形成的卡座。
The seats formed by the round wood grids.

福马桃花坞 Fuma Taohuawu

设计:宋微建、李鹏 / 面积:230 m² / 主要材料:旧墙砖、旧老式窗框、白色涂料、多层板 / 坐落地点:苏州 / 工程造价:15万元 / 完工时间:2009年3月 / 摄影:宋微建

Design:Song Weijian、Li peng / Area:230 m² / Main Materials:brick wall、window frame、white coating、continuous layer board / Location:Suzhou / Project Cost:RMB 150 000 / Finish Time:Mar,2009 / Photography:Song Weijian

项目改造前是一处上世纪七八十年代防疫站办公用房，设计师保留最基本的承重结构，其余部分均被拆除。所有拆下来的建筑材料都被回收再利用，最大限度地减少了建筑垃圾的产生和资源的浪费。从旧木市场收购的木扇门框在这里焕发出二次生机，室内的家具也由旧木组装而成，并且不用钉子和任何粘结剂，既环保又独具风格。考虑到坚固和耐用性，对这些旧材料做了技术上的加工处理，如砌筑红砖的水泥标号就比原来高了几个等级。

室内灯光的设置，尽量隐藏光源，以避免刺眼。在大办公区的桌子上方是一块很大的布幔，像一个倒置的创意表演场，而它更重要的功能是制造柔和的照明效果。窗户上安装竹帘，在日光过强的时候起到一定的过滤作用，但又不阻断窗外的精致。另外，由于地面和墙面都是刚性的棱角，采用绿植、布幔和珠帘的形式可以中和这种硬的介质，形成材料的对比和协调。门窗保持原样，形成通风和自然采光系统。空间格局颇有几分江南园林的气质，从门厅到总台，再进入办公区，曲直有章。大小办公区间采用花格，气韵通畅又互不干涉。

Before the modification, it was an office building for the epidemic prevention station in the 1970s and 1980s. The designer has reserved its most basic load-carrying members while the other parts are all removed. In consideration of the solidness and durability, these old materials have been processed technically. For example, the grade of the cement for red bricks laying has been several grades higher than the original one.

As for the setting of the interior light, the light sources shall be hidden to the utmost in case of dazzling. One cloth curtain is placed on the table in the big office area to create a soft lighting effect. The bamboo curtain is installed for the window in order to take filtration function when the light is too strong. Furthermore, as the floor are the wall all have rigid edges, the green plants, cloth curtains and bead curtains can be applied to neutralize such hard media and form a contrast between the materials. The doors and windows are kept as original to form the ventilation and natural day-lighting system. The space is arranged in an orderly curving and straight way from the hall to reception desk and then office area. The big and small office areas are divided by the grating so that air is smooth without obstruction and interference with each other.

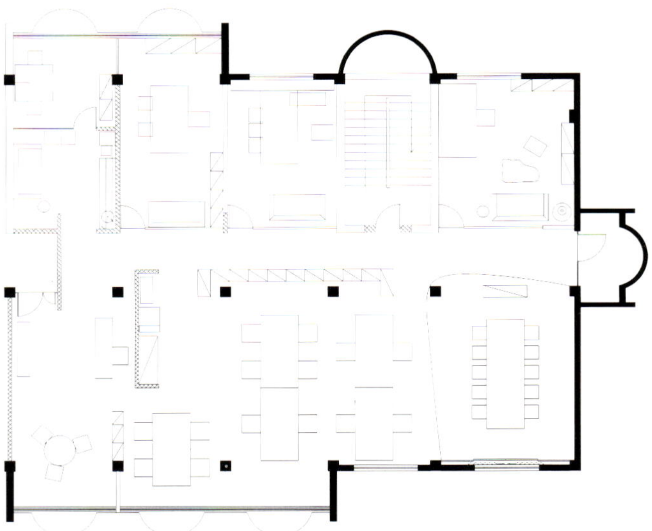

材料的对比和协调。
The contrast between the materials looks harmonious.

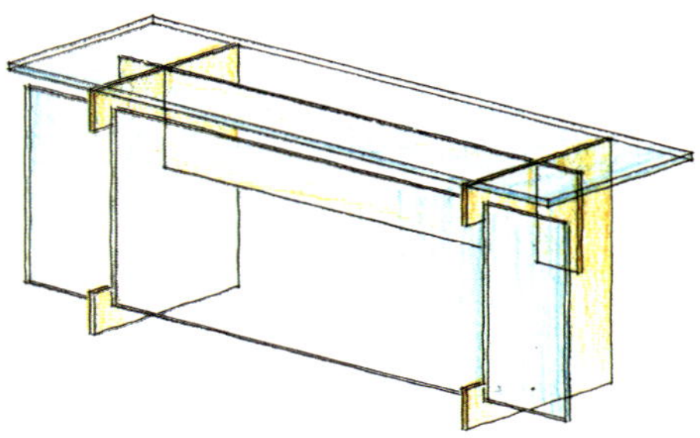
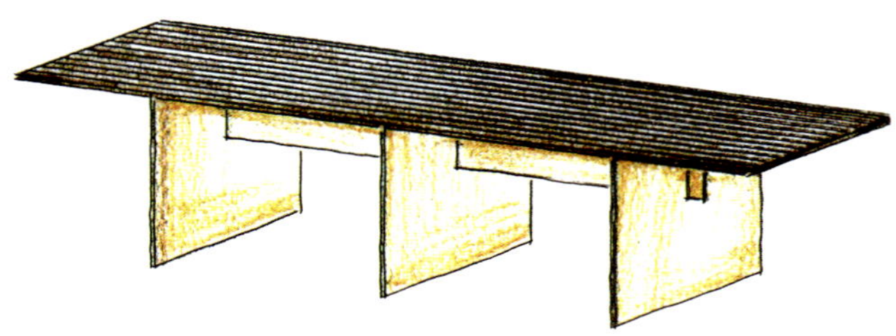

会议区。

The conference area.

过道。

The corridor.

服务台。

The reception desk.

木格样式。

The wood pane pattern.

空间布局颇具几分江南园林的气质。

The space is arranged with somewhat a charm of gardens on the south bank of the Yangtse River.

半木明舍家具展厅 Bamoo Mingshe Furniture Exhibition Hall

设计:吕永中 / 面积:500 m² / 主要材料:木材、钢材 / 坐落地点:上海 / 工程造价:50万元 / 完工时间:2008年10月 / 摄影:莫尚勤

Design:Lv Yongzhong / Area:500 m² / Main Materials:wood、steel / Location:Shanghai / Project Cost:RMB 500 000 / Finish Time:Oct,2008 / Photography:Steve Mok

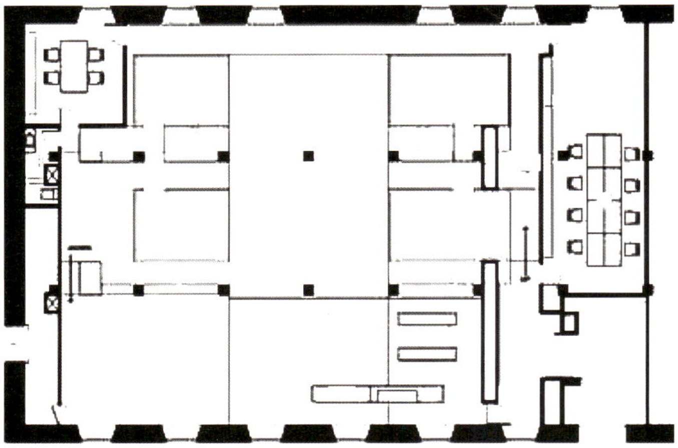

半木明舍位于上海中山公园靠近苏州河的湖丝栈创意园，为半木原创家具提供展示研究的空间。总面积近500 m²。基地是有着百年历史的砖木结构仓库，前身是用于湖州蚕丝的加工和储存工厂。斑驳的近70cm厚的清水砖墙体有如城墙般坚固与封闭，硕大的木结构梁柱记录着曾经岁月的沧桑，幽暗阴冷空间带着历史的烙印，有些古旧的沉重。

设计师首先思考的问题是“轻”与“重”的问题。空间设计重心并不仅在于具体的“象”表现本身，而是在于超越“象”之后获得的更加自由的感受和领悟，并由此展现的一个可拓展的意蕴空间，此案在这方面进行了一些探索。

推开这扇门，由现实的纷杂进入黑色通道，梦境的帷幕慢慢打开。你无法直接进入大厅，得穿过狭窄的20几米长的“一人巷”，射灯把爬在墙上的家具模型的影子拉得修长。设计师在设计之前曾多次往来此地以模拟体验并在动态中把握空间感受。这不再是用笔在工作室完成的设计，更像是用脚步走出来的设计。

整个基地平面呈长方形，三面是清水砖墙围合墙体，质感丰富，仅用清洁修补方法予之整体保留。依据柱阵沿东西方向划为两大功能部分，右边部分是隐藏的办公室，左边部分是展示空间。这两大功能块拉出狭长通道成为“一人巷”。按“田”字布局的展示大厅中，四个半透明玻璃围成有“亭”的意味，可以定义为有主体的展示空间；间隔两条水平镜面，反射室内空间制造出奇幻效果的如“水榭”；中间部分用老房子拆下来的旧地板重新加工铺装地台形成开放式中央舞台。设计师欲创造一个有戏剧性效果的空间，多条小径相互交叉，与“一人巷”入口不同，进入大厅后无固定流线方向显示，走走停停，有时会迷失，有时会相聚。在这里家具是舞台的主角；同时洁净中性的界面处理也能轻易转换功能用于其他表演展览之用。值得一提的是灯光设计，更多关注影子变化，并多处使用智能感应照明，既节约能源，又能营造不断变化的光影效果。

或许，半木如设计师所言：“取半舍满”正是体现了人生的态度及“朴拙美好”的始终信仰。

The base of Banmu Mingshe is a brick—timber structure warehouse with a history of over one hundred years and was formerly the fabrication and storage factory for silk in Huzhou.

The importance of space design is to obtain the more free feelings after surpassing the “shapes” besides the specific demonstration of itself to exhibit an implicative and expandable space.

The base plane occurs as a rectangular, enclosed by dry brick walls in three sides and reserved entirely in way of cleaning and repairing. It is divided into two main functional areas from the east to the west along the columns: the hidden office in the right and exhibition hall in the left. In the exhibition hall, four pieces of translucent glass are enclosed into a “pavilion”; two horizontal mirrors are reflecting the interior space to create a fanciful effect; in the middle, an opening central stage is set up by laying the platform with the re-fabricated used floor boards which were dismantled from the old house to create a space of dramatic effects. As for the light design, more intelligent inductive lighting is applied to both save the power and develop a changeable light-and-shadow effect.

一个可拓展的意蕴空间。
A developable meaningful space.
展厅一角。
One corner of the exhibition hall.
展厅内部。
The interior exhibition hall.

展厅一角。
One corner of the exhibition hall.
斑驳的清水砖墙体。
The blocky fair-faced brick wall.
展示家具一。
Exhibited Furniture 1.
展示家具二。
Exhibited Furniture 2.

展示家具三。
Exhibited Furniture 3.
展示家具四。
Exhibited Furniture 4.
展示家具五。
Exhibited Furniture 5.
展示家具六。
Exhibited Furniture 6.

小海发廊 SHOWHAIR

设计:宋微建、郭纯 / 面积:140 m² / 主要材料:亚克力、古铜、皮毛、马赛克、肌理漆 / 坐落地点:上海 / 工程造价:45万元 / 完工时间:2008年7月

Design:Song Weijian、Guo chun / Area:140 m² / Main Materials:PMMA、bronze、fur、mosaic、paint / Location:Shanghai / Project Cost:RMB 450 000 / Finish Time:Jul,2008

在这个时尚造型会所的设计上，混搭风格就是设计核心。

店铺是一栋有着一百年历史的西班牙建筑。欧式的线角、破旧的门窗、残缺不一的磨纹玻璃。虽然房间里破烂不堪，但是却磨灭不掉历史留下来的痕迹和味道。

设计师保持内部结构的完整性，只对平面做些许改动，从而满足功能上的一些需求。在材质的选择上，一种为现代前卫，一种为古朴怀旧，目的也是为了更好的加强对比，让现代的时尚和历史的印记发生激烈的碰撞。

亚克力大面积的使用是设计中的一个亮点，在窗口、柜子、灯罩、窗帘盒的设计上都分别运用了12mm厚的透明亚克力，老的结构和旧构件被更好地显露出来。

镜子在设计中也是重中之重，选用古铜为材料，穿插了皮毛的使用，这样不但打破空间生硬的感觉，还体现出雍容华贵的感觉。

除了会所的设计外，这里边还有一间经理的办公室设计，界面上只做简单处理，在软装配饰上下了很大的工夫，整体以黑白灰做为基调，家具和装饰品做为空间的主角。透明的亚克力椅子、皮制沙发、铜雕塑、装饰画，打造轻松舒适且具有品位的办公空间。

In the design of this fashionably modeled club, the mixed style is the core of the design. The store is a Spanish building with a history of one hundred years. Although the interior rooms are broken, the races and smells that the history leave behind can not be worn away.

The designer has kept the integrity of the interior structure and only made slight changes on the plane. With regard to the material selection, one is modern and fashionable while the other is antique, simple and reminiscent.

The application of the large area of acrylic is a shining point of this design. The transparent acrylic has been applied to the windows, cabinets and curtain boxes. The mirrors are also the highlight in the key points in the design. The antique bronzes are selected as material with alternation of the leathers and furs, which not only breaks through the rigidity of the space, but also shows a natural, graceful and luxurious atmosphere.

With regard to the design of manager office, the interfaces are just processed simply while great efforts have been taken to the soft decoration. Black and white is the entire basic color and the furniture and decorations play a leading role in the space to create a relaxing, comfortable and tasty office space.

卫生间盥洗台镶着金色的边。
The washing table in the washroom is inlayed with the golden edges.
接待区前卫与时尚并存。
The pioneering and fashion coexist in the reception area.
马赛克装点的前台呈现低调的奢华。
The reception desk decorated by the mosaic has delivered a low-profiled luxury.

SHOWHAiR™
NEW LIFE STYLE
SHOWHAiR™

破旧的门窗渲染历史留下来的痕迹和味道。

The broken door and window are exaggerating the traces and smells left behind the history.

古旧的味道充斥着这间小屋。

This small room is full of antique and old smells.

镜子在设计中选用古铜为材料，穿插了皮毛的使用。

The mirror in this design has applied to the old bronze as the material, alternated with the furs.

草流行中药养生馆 Chinese Herb Medication Center

设计:邓琨艳 / 面积:440 m² / 主要材料:老洋松木材、青砖、玻璃、山水白大理石、望砖 / 坐落地点:上海 / 工程造价:100万元 / 完工时间:2009年3月 / 摄影:莫尚勤

Design:Deng Kunyan / Area:440 m² / Main Materials:old pine、blue brick、glass、marble、shedthins tile / Location:Shanghai / Project Cost:RMB 1 000 000 / Finish Time:Mar,2009 / Photography:Steve Mok

这是原先上海通用电气20世纪30年代旧工厂的生活用区的旧环境。对于这座新改装再利用建筑的规划与环境的空间尺度，除了尽量保护，也希望做到让整体环境更和谐。在不突兀的情况下，让这座中药养生馆突显内涵。

基于旧建筑再利用的思考，拆除一部分完全不堪使用的构造物。然后在原有的基础上，因着新的使用功能，长出当下该有的建筑物。材料的选用基于设计师一直秉承的建筑旧材料再利用的环保观念。

It used to be an old environment of the living quarters of the old Shanghai General Electrical factory in the 30' s of the last century. As for the planning and the ambient space size of the newly renovated building, it intends to make the whole environment more harmonious besides the maximum protection so that this Chinese herb medication center can prominently show its meaning naturally.

Based on the consideration for the reutilization of the old building, the parts of the structures that can not be used any more were dismantled and the appropriate buildings are built up on the original foundations due to the new functions. The selection of the materials is based on the environmental protection philosophy of recycling the used construction materials that the designer has been keeping.

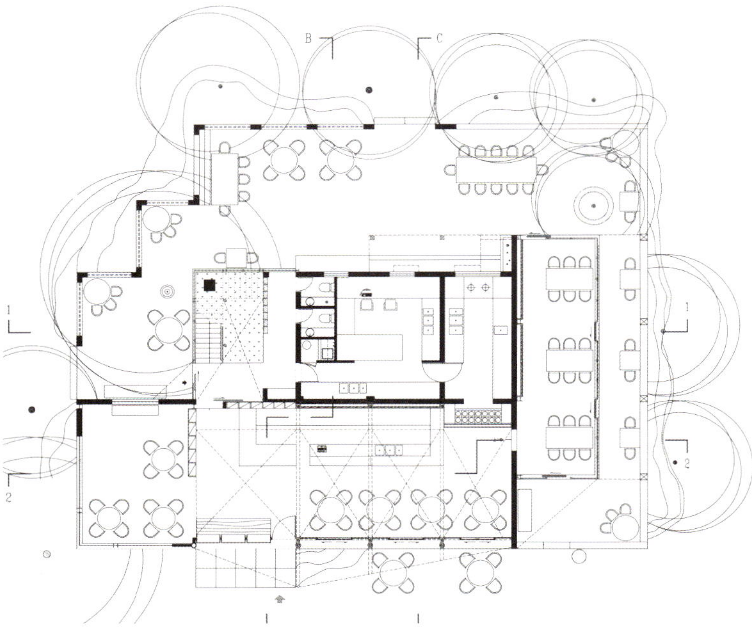

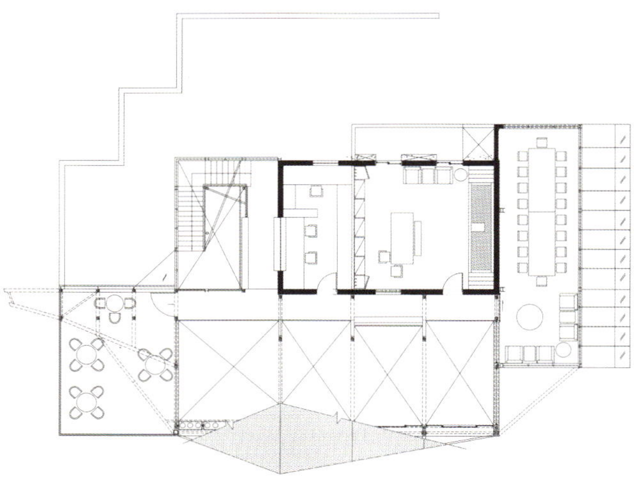

绿树红花掩映的世外桃源。

A Land of Peach Blossoms protected by green trees and red flowers.

老建筑原本的材质被保留。

The original materials of the old building are reserved.

服务台的摆设呼应中草药的主题。

The arrangement of the service desk corresponds to the theme of Chinese herbal medicine.

瑞亚·臻品酒店（金桥路店）

RIA-Perfection Hotel (jing-qiao road's)

设计单位:美国PDA室内设计事务所 / 设计:裴晓军 / 面积:6 000 m² / 主要材料:诺娃米黄石、玫瑰木、壁纸、布艺、钢材 / 坐落地点:上海 / 工程造价:900万元 / 完工时间:2008年5月 / 摄影:贾方

Design Unit:USA PDA DESIGN / Design:Pei Xiaojun / Area:6 000 m² / Main Materials:rosewood、wallpaper、cloth、steel / Location:Shanghai / Project Cost:RMB 9 000 000 / Finish Time:May,2008 / Photography:Jia fang

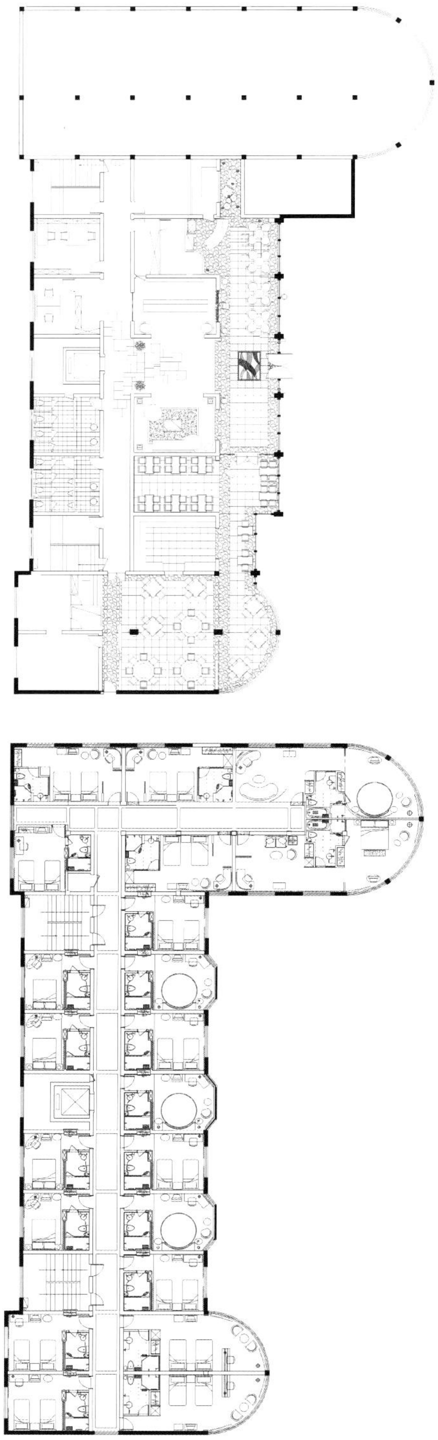

过廊。
The corridor.

把一个80年代的镇政府写字楼变成一个精品酒店面临多重制约，理性与感性之间发生不断的转换和游离。设计的投入分配不同寻常，它演变的过程精彩于结果。这个探索对于旧建筑的商业开发具有启发和示范意义。建筑六层的承重墙结构，4 m间距一字排开，这些“小房子”找不到大堂的位置，它们相互封闭，其他必备的共享公共空间似乎成为不可能。还有垂直设备井、消防系统的重建等等，未来会成为怎样一种酒店功能格局？

·建筑改造功能空间的重建

整合内部空间的方法是与建筑外立面同步进行。从建筑外墙面向院内增扩4.5 m，将这个外立面的门窗洞打开加固。虚张声势通过外加的钢结构阳光厅廊赢得一箭多雕之功效：二进式大堂丰富互补；非廊非堂室内外彼此交流，强调了人与自然的融合；365天的阴与阳改善了原建筑的采光不足；在这个廊内，堂吧和餐厅的位置自然形成、顺理成章；改变了建筑外立面的形态增加了建筑的体量和过度层次，酒店识别性增强；扩大的这部分使建筑空间发生了质变，阴阳、虚实结合得体。

·室内空间表现与文化的重塑

这个时期是对建筑和功能空间的整合结果进行再创造和深化。其中包括空间的表现、生活的肌理、文化的植入、动静空间的分配等。通过特有的简练装置方法注入思想和生命，汇入业主规定的地中海风情；虽然是一条二维的鱼，它的出现已“干扰”这个区域的三维空间，使僵化了的建筑空间活跃起来；这里尽可能满足酒店对必备公共功能空间的要求，强化客房使用的舒适度和魅力。

The designer first should face architecture, a new hotel or reconstruction of a hotel; reconstructing the government office building in 1980s to a new Top-quality hotel, we need verify what the chance of this reconstruction, and what conditions this building could offer and create? Last forward a method, and achieve the target.

The reconstruction face many restrictions, make the designer transform and dissociate from sense to sensibility. The devotion and distribution of design are unusual; the process of evolution is more fascinating than the result. This reform has enlightenment and demonstration significance to commercial exploitation of the old structure.

The structure has six floors, 6000 square meter bearing wall, room space is 4 meter, and stand in one line, these small rooms could not find the hall, they are completely sealed each other, having share space is luxurious and impossible. Also we could reconstruct the vertical equipment well and fire prevention system. What the function and layout will the reconstructed hotel display in future?

Reconstruction of function space

Integrating the inside space must synchronize with reconstructing external facade of the architecture. Spreading inside 4.5 meter from external facade, opening and reinforcing these windows and doors of the external facade, additive steel structure sunny porch could bring multiple effects: stepping in the hall through the sunny porch, these two areas exchange themselves, stressing the human blend in with the nature; the luminous porch improving the daylight for the structure; the bar and restaurant are ready-made in this porch; reconstructing external facade of the structure could increase dimensions and transition layers, improving the identification of the hotel; extending area make the architecture qualitative change, combining yin with yang, false with true propriety.

Display of inside space and reengineering the culture

This time recreating and deepening the effect of integrating the architecture and function space. Including display of space, details of live and culture involved, dynamic and quiet distributed and so on. Breathing into thought and life through the concise style, flowing into Mediterranean Civilization simultaneously according to the proprietor demand; this two-dimensional fish interfere and control the three-dimensional field of this area, making the rigid architecture space lively; trying the best to fill the requirement of necessary public space, insist on comfort and magic of the room. Because of the partition of architecture, the structure space enrich style of room, to compound different space power, includes the smallest guest room in the world.

地中海风情。
The Mediterranean charm.
艺术品。
The artwork.
休息区。
The rest area.
休息区。
The rest area.

客房过道。

The corridor of guest rooms.

大床间。

The king size room.

才茂街厂区改建项目 The reconstruction project of Caimao street

设计单位:IADC涞澳设计公司 / 设计:张成喆 / 面积:1000 m^2 / 主要材料:旧木、砖、铜板、镜面钢、特殊工艺 / 坐落地点:武汉 / 完工时间:2009年9月 / 摄影:蔡峰

Design Unit:IADC DESIGNERS LTD. / Design:Elio Zhang / Area:1000 m^2 / Main Materials:old wood、brick、copperplate、mirror surface steel、special technology / Location:Wuhan / Finish Time:Sep,2009 / Photography:Cai feng

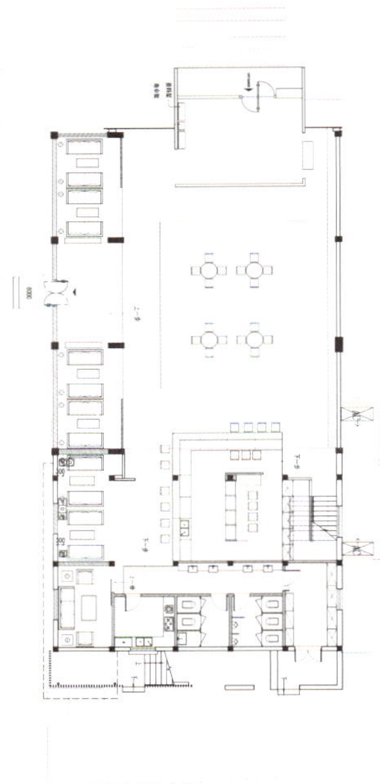

本案原址是一座混凝土工场，规划中保留了原厂址部分结构和厂房。作为新项目的一个组成部分，新增建筑和景观围绕厂区结构进行规划和改造。新旧之间的关联成为项目的一个重要特点，作为建筑和景观的最终延伸。接待中心的设计成为项目中重要的组成部分。

尊重原建筑构造的特点，部分建材更取之于项目中拆除和废弃的材料，如天花的旧木和地面的红砖，均是基地中的废弃材料，墙面也按照原建筑的做法进行复原，甚至保留了厂房建筑中的很多细节。

以工业铜板为材质的方盒体构造形成建筑入口，30 m长的影像墙面连接了入口和主要空间。摄影作品展示了基地的前世今生，成为空间中的焦点。空间顶部悬着的金属片构造仿佛浮云般反射了光线，在不同光线下形成有趣的光影变化。端头的酒吧区营造了舒适的氛围，吧台由一整片混凝土浇制而成。

室外景观和整个环境取得有机的联系，在空间和材质等方面做了许多研究。尊重历史建筑的同时，注入新的功能和设计语言，使这一有50年历史的厂区焕发出新的生机与活力。

The original address of the case is a concrete work site, the part structure and factory room of the original factory address are maintained in the program.As the composed part of the new project, the structure of the added archtecture and landscape around the factory area are undertaken the program and reconstruction. The link between the new and old is the important feature in the project and as the final extension of the building and landscape.the design of the reception center becomes the important composed part of the project.

With expection to the features of the original building, part construction materials derived form the dismantled and discarded materials of the project, such as the old wood of the ceiling and the red breaks of the floor, both are the discarded materials of the base.The wall surface are recovered according to the previous construction methoed, and even maintain a lot of details of the previous factory construction.

Square box structure made from the industry copper plate form the the entrance of architecture, the image wall surface for 30 meters connect the entrance and the main space. The image work displays the past and the present, become the focus of the space.The sheet metal structure hanged in the ceiling reflect the light like the drifting clouds, form the interesting morphing lights and shadows under the vious light.

The relationship between the new and old materials.The bars districts at the end make a convenient atmosphere, the bar dests are made form the concrete casting.

The organic link between the indoor landscape and the whole surroundings, plentiful research about the space and materials are made, with respect to the historic building, meanwhile, bringing the new function and design language make the factory area with 50 years history show new vigor and vitality.

建筑入口。
Entrance of architecture.
原厂房的许多构造被保留下来。
A large number of structure of the previous factory are maintained.
天花的旧木和地面的红砖，均是基地中的废弃材料。
Both of the old wood of the ceiling and the red break of the floor are the discarded materials.

以工业铜板为材质的方盒体构造形成建筑入口。
Square box structure made from the industry copper plate form the the entrance of architecture.
新旧材质之间的关联。
The relationship between the new and old materials.
红砖和实木的搭配返璞归真。
Match of red break and solid wood recover the simple and come to the nature.
端头的酒吧区营造了舒适的氛围。
The bars districts at the end make a convenient atmosphere.
狭长的盥洗区。
Long and narrow bathroom.

顶部悬着的金属片构造在不同光线下形成有趣的光影变化。

The sheet metal structure hanged in the ceiling form the interesting morphing lights and shadows under the various lignt.